Praise for Richard Rhodes's

Scientist

"It has been an honor to know Ed Wilson. His life and work have inspired so many—scientist and layperson, alike. Richard Rhodes's *Scientist* is a wonderful introduction to one of the great thinkers and observers of our age." —Paul Simon

"An impressive account of one of the twentieth century's most prominent biologists, for whom the natural world is 'a sanctuary and a realm of boundless adventure; the fewer the people in it, the better.'"
—*The New York Times Book Review*

"Wilson's life and substantial accomplishments—many have called him the 'natural heir' to Darwin—are ripe topics for exploration, and particularly important as we continue to confront the climate crisis's effects on biodiversity." —Literary Hub

"Pulitzer winner Rhodes (*The Making of the Atomic Bomb*) does justice to 'one of the . . . greatest biologists of the twentieth century' in this brilliant biography. . . . Rhodes depicts Wilson as a tireless field scientist at a time when the general belief was that the future of biological discoveries was in the laboratory, and as a proponent who popularized sociobiology, and as a Pulitzer winner for his books *The Ants* and *On Human Nature*. The author leaves no doubt as to Wilson's broad impact on science and the public's perceptions of nature, without ever veering into hagiography. This is a must-read."
—*Publishers Weekly* (starred review)

"Esteemed biographer and historian Rhodes warmly portrays Wilson as an ambitious and accomplished biologist, a passionate and influential advocate for identifying all life-forms and preserving half of Earth as natural habitat, and a prolific, Pulitzer Prize–winning writer. . . . Rhodes's biography makes a fine companion to Wilson's *Tales from the Ant World* (2020)." —*Booklist* (starred review)

"Rhodes (who won a Pulitzer for *The Making of the Atomic Bomb*) devotes as much time to Wilson's remarkable life as to his remarkable achievements as a biologist, making this biography a joy to read." —*The Washington Post*

"*Scientist* offers a chronological guide to Mr. Wilson's proliferating research interests, providing succinct, nuanced summaries of some of his major insights, enriched by frequent forays into the history of modern biology. Among the most delightful sections in *Scientist* is Mr. Rhodes's reconstruction of the battle between Mr. Wilson and his Harvard colleague James Watson, codiscoverer of the double helix structure of DNA, who mocked field scientists as 'stamp collectors.'" —*The Wall Street Journal*

RICHARD RHODES

Scientist

Richard Rhodes is the author of twenty-four books, including *The Making of the Atomic Bomb*, which won the Pulitzer Prize in nonfiction, the National Book Award, and the National Book Critics Circle Award. He graduated from Yale University and has received fellowships from the Ford Foundation, the National Endowment for the Arts, the John Simon Guggenheim Memorial Foundation, and the Alfred P. Sloan Foundation. He has been a visiting scholar at Harvard, MIT, and Stanford, and a host and correspondent for documentaries on American public television. He lives in Seattle with his wife, Dr. Ginger Rhodes, a clinical psychologist.

ALSO BY RICHARD RHODES

Energy

Hell and Good Company

Hedy's Folly

The Twilight of the Bombs

Arsenals of Folly

John James Audubon

Masters of Death

Why They Kill

Visions of Technology

Deadly Feasts

Trying to Get Some Dignity
(with Ginger Rhodes)

Dark Sun

How to Write

Nuclear Renewal

Making Love

A Hole in the World

The Making of the Atomic Bomb

Looking for America

Sons of Earth

The Last Safari

Holy Secrets

The Ungodly

The Inland Ground

Scientist

Scientist

E. O. WILSON: A Life in Nature

RICHARD RHODES

Vintage Books
A Division of Penguin Random House LLC
New York

FIRST VINTAGE BOOKS EDITION 2023

Originally published in hardcover in the United States by Doubleday, a division of Penguin Random House LLC, New York, in 2021.

Grateful acknowledgment is made to Harvard University Press for permission to reprint previously published material from the following:

The Insect Societies by Edward O. Wilson, Cambridge, Mass.: The Belknap Press of Harvard University Press. Copyright © 1971 by the President and Fellows of Harvard College. Used by permission. All rights reserved.

Sociobiology: The New Synthesis by Edward O. Wilson, Cambridge, Mass.: The Belknap Press of Harvard University Press. Copyright © 1975 by the President and Fellows of Harvard College. Used by permission. All rights reserved.

On Human Nature by Edward O. Wilson, Cambridge, Mass.: The Belknap Press of Harvard University Press. Copyright © 1978 by the President and Fellows of Harvard College. Used by permission. All rights reserved.

The Library of Congress has cataloged the Doubleday edition as follows:
Names: Rhodes, Richard, [date] author.
Title: Scientist : E. O. Wilson : a life in nature / Richard Rhodes.
Description: First edition. | New York : Doubleday, 2021. |
Includes bibliographical references and index.
Identifiers: LCCN 2021944184
Subjects: Wilson, Edward O. | Naturalists—United States—Biography. |
Entomologists—United States—Biography.
Classification: LCC QH31.W64 R46 2021 | DDC 508.092 B—dc23
LC record available at https://lccn.loc.gov/2021944184

Vintage Books Trade Paperback ISBN: 978-1-9848-9835-7
eBook ISBN: 978-0-385-54556-3

Author photograph © Nancy Warner
Book design by Cassandra J. Pappas

vintagebooks.com

Printed in the United States of America
10 9 8 7 6 5 4 3 2 1

For Ginger

A grant from the Alfred P. Sloan Foundation supported the research and writing of this book.

Nothing in biology makes sense except in the light of evolution.

—THEODOSIUS DOBZHANSKY

Contents

Scientist

1

Specimen Days

Museum of Comparative Zoology
at Harvard College
Cambridge 38, Massachusetts
Director's Room

November 2, 1954

To whom it may concern:
This will introduce Mr. Edward O. Wilson, a junior fellow in Harvard University. Mr. Wilson is traveling to Australia, Ceylon, New Guinea and New Caledonia for the purpose of collecting scientific specimens for the Museum of Comparative Zoology. Any assistance or advice you may be able to give him will be greatly appreciated.

Very truly yours,
Alfred S. Romer
Director

FINALLY, Ed Wilson was on his way, twenty-five years old, tall and lanky, the upper range of his hearing gone since his teens, his right eye ruined in a childhood accident: half deaf and half blind. Outwardly, he was a polite, soft-spoken product of Gulf Coast Alabama, the first in his family to graduate from college. But behind the well-mannered finish he was as tough as nails, as bright as the evening star, and no man's fool. He would become one of the half-dozen greatest biologists of the twentieth century. In the new century now advancing, he would lead the charge to save what's left of wilderness—half the Earth, he said—not only for the experience of wilderness but also for the millions of species large and small, many of them not yet even named, in danger of going extinct, forever. If they did, he taught, they would take with them their supporting strands of the great web of life, unraveling the world. Trees can be replaced; species, having evolved into being across millions of years, are irreplaceable.

For now, a fresh-minted Ph.D., just setting out, his lifework before him, young Wilson was bound for the South Pacific to collect ants. Entomology was his field—insect biology—and ants were his specialty. No one had ever systematically collected ants across the vast sweep of the South Pacific. For ant specimens, many Pacific islands had never been explored.

The new frontier in 1954, after James Watson and Francis Crick's great 1953 discovery, was the structure and function of DNA. Biologists everywhere had run to their labs to inform their science with chemistry and physics. Wilson was deliberately running the other way. "If a subject is already receiving a great deal of attention," he explained his strategy later, "if it has a glamorous aura, if its practitioners are prizewinners who receive large grants, stay away from that subject."

Wilson was an explorer at heart, had been since he was a small boy. Rather than focus his work on one species, as most of his peers would do, he preferred to break new ground, find the big nuggets, assay them, and move on. If he'd started his life as an only child in the Gulf Coast South, chasing down snakes and butterflies, he'd arrived at Harvard for graduate study in 1951 as a certified prodigy. In a vacant lot in Mobile, Alabama, when he was only thirteen years old, he'd been the first collector in the United States to spot the invasion of the pestilential red imported fire ant, *Solenopsis invicta*, transported from Argentina as a ship stowaway. Seven years later, as an undergraduate at the University of Alabama, he'd published his first scientific paper.

Avoiding the crowd was a risky strategy—one that would reward him repeatedly across a long, successful career, but would also vex him with major challenges. "Take a subject instead that interests you and looks promising," his advice continues, "and where established experts are not yet conspicuously competing with one another. . . . You may feel lonely and insecure in your first endeavors, but, all other things being equal, your best chance to make your mark and to experience the thrill of discovery will be there." The advice was vintage Wilson, potentially beneficial in equal measure to the curious boy and the ambitious adult.

A Harvard Junior Fellowship stood behind his one-man expedition to the South Pacific on behalf of the Harvard Museum of Comparative Zoology. The exclusive Society of Fellows, founded at Harvard in 1933, annually awarded a dozen exceptional young scholars three years of support for any research or study they chose to undertake. Across the decades, the ranks of Harvard Senior and Junior Fellows have included such well-known leaders as the presidential adviser McGeorge Bundy, the historian Crane Brinton, the Nobel-laureate physicist and Manhattan Project veteran Norman Ramsey, the strategic analyst Daniel Ellsberg, the economist Carl Kaysen, the transistor co-inventor John Bardeen, the artificial-intelligence pioneer Marvin Minsky, the geographer and historian Jared Diamond, the behavioral psychologist B. F. Skinner, and the U.S. poet laureate Donald Hall. In 1954, the Society added Ed Wilson's name to its list.

Collegial dinners every Monday night in term time introduced Wilson to such visiting stars as the charismatic J. Robert Oppenheimer, a high-school hero of Wilson's at the end of World War II, when Oppenheimer emerged as the so-called father of the atomic bomb. "Promethean intellect triumphant," Wilson would recall his assessment of the famous physicist, "master of arcane knowledge that had tamed for human use the most powerful force in nature." Junior

Fellows were expected to attend such weekly dinners as well as twice-a-week luncheons with their peers. Wilson would miss a share of both in his nine-month Pacific expedition.

He would miss something—someone—else as well, a far more intensely personal deprivation. He and a young Boston native, Irene Kelley, dark-haired and pretty, had become engaged not long before, after a yearlong courtship. "The first time I ever saw you," he would remember and write her from New Guinea in March, "you came down the steps from upstairs at Shirley Hayes. In the back of my mind I said, 'What a pretty girl,' but also, 'not my type, probably a party girl of the first water, being rushed by various dancing dons of the cocktail-lounge set." How wrong he was, he wrote Irene, "how exciting to me our courtship was, and how deeply satisfying to me when I began to sense your real qualities and we began to fall in love."

Irene was his first love. He had earned both bachelor's and master's degrees in only four years at the University of Alabama, carrying heavy course loads, studying through the summers, and there had been no time for dating. "My father was deteriorating rapidly," he recalled. "I could tell that I might not have any support whatever." So he had rushed through, and only now, with a Junior Fellowship in his pocket, had he begun to think of life beyond study and research. "I knew that I'd better start meeting some of the young women and developing a fuller life as an adult. In those days, the early fifties, the way you dated a young woman—or at least this was true in the South, where I'd come from—when you invited a young lady out frequently, you invited her to dinner and some kind of after-dinner entertainment, such as dancing. So I thought, I'd better learn how to dance."

He checked around Cambridge. "There was a very respectable-looking dance studio on Commonwealth Avenue which offered lessons in dancing, ordinary ballroom dancing. That's where I started my lessons." He was far from a natural. After a few lessons, preoccupied with study, he dropped out. But he was never a quitter. After a time, he began dancing lessons again. "And then I met this vision. Irene. She really was a beautiful young woman, and very gracious, even by Southern standards. I had just returned from Cuba and Mexico—my first fieldwork as a Junior Fellow. I was poorly dressed, and I was clumsy. She did the best she could with me. I just wasn't cut out for dancing. I couldn't even learn the fox-trot."

Though Irene wasn't a college graduate, she was smart and competent, both at dancing and in her day job in the Harvard University admissions office. She taught dancing in the evening for a reason Wilson immediately admired: to help support her family, which included a younger sister who was mentally ill.

"So I invited her out and we did not go dancing," Wilson recalled. "We did everything else. Fairly quickly, we became engaged. She had wonderful health at that time, and we walked practically every street in downtown Boston. We went to virtually every restaurant. We made exploring restaurants our hobby. And in this way, we developed an idyllic romance."

Then, in his second year as a Junior Fellow, at about the time he and Irene were planning to marry, he was invited to collect ants in the South Pacific for the Harvard Museum. Going away for a year might have soured their romance. "I'd dreamed of doing this," Wilson told me, recalling his childhood expectations. "I realized it could be a tremendous personal experience for me—to be a pioneer, the first to go into areas where ants had never been collected before. I explained all this to Irene. We were engaged, but I told her, 'I really need to go.' It was something like a soldier leaving for war. I explained what an extraordinary advantage I was being offered, doing original work in a completely unstudied part of the world. And she said: 'Go.' And go I did, with each of us pledged to write each other every day."

The two said goodbye at Boston's Logan Airport on 26 November 1954, the day after Thanksgiving, both of them heart-stricken at the long separation that opened before them. "I am proud that you didn't cry," Wilson wrote her later that day from Louisville, where he had stopped over to visit his mother, "but I want you to know that it was the most painful thing I had ever experienced." He didn't cry, either, not when they separated, but as soon as he boarded he started "crying like a baby," he told her:

> It isn't very manful of me to behave like this, I know, but it is really the way I feel, and the way I believe and hope you feel. I love you more deeply than I can understand. This is the first and the last time I will ever fall in love like this, and, believe me, it is the last time we will ever part the way we did [today].

Three days later, Wilson was in Honolulu, marveling at the "incredible paradise" after wintertime Boston, and waiting for a flight to Fiji. With a stop to refuel on Kanton Island, a mid-Pacific curl of atoll with a six-thousand-foot World War II runway, his flight crossed the International Date Line and the equator and arrived at Viti Levu, the largest of Fiji's more than three hundred islands, at noon on the first day of December.

"Never before or afterward in my life," Wilson wrote later of that arrival, "have I felt such a surge of high expectation—of pure exhilaration—as in those few minutes. . . . I carried no high-technology instruments, only a hand lens, forceps, specimen vials, notebooks, quinine, sulfanilamide, youth, desire, and unbounded hope." His real high-tech instrument was his brain, his heart its engine. Eager to begin, he hired a car and driver to carry him directly into the island interior, where, in lush forest, he spent the last hour of the day collecting.

"I am really in a foreign country now," he wrote Irene early the next morning, before heading back to the forest for the day. The Fijians, he teased, "gave up cannibalism some time ago," but those in the interior "still live rather primitively in grass huts." A man he had met who had known one of the last cannibals on the island had passed on the Fijian's assessment that "human flesh was salty, not as tasty as pig."

Wilson collected that day in mountain forest in hard rain, climbing steep slopes through tangled undergrowth, and returned to his hotel "soaked and dog-tired." Fiji had been depleted of wild nature by logging and settlement and previously studied in depth by his predecessor William M. Mann; he wouldn't linger there. With the autograph of the chief of the Raki-Raki tribe—well-educated, in a torn T-shirt—mailed off to Irene for safekeeping, the young biologist departed Fiji by flying boat for New Caledonia the following morning. "I want to share everything I gain on this trip with you," he wrote his fiancée before he left, "and I sort of feel that I am undertaking every new adventure for you as well as for myself."

James Cook, the British navigator, discovered New Caledonia in September 1774, on his second voyage of exploration. An island about the size of New Jersey, some seven thousand square miles, but long, narrow, and mountainous, it reminded Cook of his native Scotland—"Caledonia" to the Romans—and so he named it. "It will prove at least 40 or 50 leagues [120–150 miles] long," observed his onboard naturalist, Johann Forster, "& is therefore the greatest new Tropical Island we have hitherto seen." Some nine hundred miles northeast of Australia, it was in fact the third-largest island in the Pacific, after New Guinea and New Zealand. The French seized it from the British in 1853 and used it as a penal colony for both men and women until the end of the nineteenth century. During World War II, it hosted some fifty thousand Allied troops; the U.S. warships that fought the Battle of the Coral Sea in May 1942 had been harbored there.

New Caledonia would be Wilson's favorite Pacific island. French colonials and indigenes alike were friendly, and no professional had ever collected ants there. Wilson arrived on Saturday, 5 December 1954; by the next evening, he had already made two field trips, one to a valley forest outside the capital, Nouméa, a second to a nearby mountain.

Monday, 6 December, he stayed in town to go over his first day's collection. He found, to his delight, that he already had three or four new species. "My work here will no doubt have fabulous results," he wrote Irene. "I expect to double the known number of New Caledonian ants." He'd found his way as well to the New Caledonian Institut Français d'Océanie—the French Institute of Oceania—introduced himself, and worked in its entomological laboratory. New Caledonia was "relaxed and easy," he added, "with none of the British snap and discipline evident on Fiji." The place put him "in fine spirits" except for the high prices, running twice what he'd budgeted. Those he attributed to bad French management: "The reason for the high prices is that the French have a stagnated, unprogressive economy and little imagination to devote to the welfare of the colony."

He spent the next day working with a borrowed microscope in the lab at the Institut Français, preparing his new specimens, buoyed by a week's worth of letters from Irene that had arrived all in one batch that morning. He'd arranged them to read in chronological order. "It wasn't too hard setting my mind back a week as I read what you were thinking from day to day," he wrote her. "It made me very happy and carried me back to you in Boston in a very real fashion for a little while." With the last of the letters, Irene had enclosed her photograph. It showed how pretty she was, he told her, and promised her he would carry it with him everywhere and "show it to every Frenchman, lizard, and butterfly I can buttonhole." He apologized for not calling her before he left the United States; he'd felt miserable, he wrote, and knew he'd have "broken down" had he called. Now, a week out, he was "feeling much better and . . . remembering all the happy hours and days we had together and forgetting the really few hours of sadness [when they separated]." Then he had to break away from letter writing—there were French scientists walking around the table where he was writing, and it made him self-conscious. He quickly signed off, "All my deepest love."

His days fell into a routine: up at 6:00 a.m.; fruit, coffee, and bread for breakfast at a small French restaurant near his edge-of-town hotel; then off to what he called a "little patch of forest" at Le Chapeau Gendarme, a peak shaped like a French policeman's cap in a ridge of mountains behind Nouméa, for a day's work observing and collecting. Reviewing his life many years later, he celebrated his good luck: "At the time I entered college only about a dozen scientists around the world were engaged full-time in the study of ants. I had struck gold before the rush began. Almost every research project I began thereafter, no matter how unsophisticated (and all were unsophisticated), yielded discoveries publishable in scientific journals." New Caledonia was only the beginning.

The highlight of his exploring there was a climb to the summit of Mount Mau, at 469 meters (about fifteen hundred feet) the third-highest peak in the country. The climb itself, he wrote Irene, was "a stiff little hike, very hot and wearisome," but at the top he found cloud forest, "a weird and wonderful world of little twisted trees covered with thick moss and epiphytes [air plants], and various weird little plants growing on the moist, mossy forest floor." Clouds limited his view—he could see no more than fifty feet in any direction—but what he saw was magical:

> It was truly a lost world, hardly ever visited by humans, in its primeval state, and utterly unlike any other kind of forest you see in the lowlands. I picked up at least one new species of ant and saw a brilliant green parrot, who sat on a branch about 15 feet away and squawked at me. . . . In the West Indies this kind of forest is called "elfin forest" because of the funny little twisted, moss-covered trees.

Wilson continued day-collecting in New Caledonia through Christmas and New Year's, adding to his list of new species and adjusting to the tropics. In Nouméa there were "absolutely no signs of Christmas decorations in the stores," he marveled. For the French, Christmas was for children, New Year's for adults. Santa Claus was supposed to arrive by French submarine, but could only muster an old tugboat; he was greeted anyway by an enthusiastic crowd of more than a thousand people, "including many children and fascinated New Caledonian natives and Indochinese and Malayans." At the American consulate's picnic on Christmas Day, everyone gathered around as the visiting young Harvard entomologist, whether sober or otherwise he doesn't say, "tried to climb a coconut tree (nobody else would dare); I got halfway up and pooped out, and came down with only a couple of scratches for my troubles." All in all, it had been "the strangest Christmas I ever hope to spend."

Taking a break at the beach one afternoon, he noticed a few of the new bikini-style swimsuits, named after the site of the first post-war U.S. atomic-bomb tests, conducted twenty-three hundred miles north, on Bikini Atoll. He hastened to assure Irene that most "colonial French are extremely conservative in dress and manner." He dived the reef, using a newfangled "snorkel" and fins, tried to catch a fish by the tail, hoped to see a shark one day, "but none of this riding-on-the-shark's-back-for-sport for me." He'd made a crowd of friends; he sent Irene for safekeeping a list of twenty-five names, ranging from three fellow entomologists, two French and one New Zealander, to the secretaries at the American consulate, to the workers at the little French restaurant where he ate his meals, to an Australian filmmaker. From their perspective, he was the exotic.

Early in the new year, investing a week on behalf of both "curiosity and opportunism," Wilson diverted to what he called "just about the most remote spot on earth which you can reach by the usual airlines," the island of Espiritu Santo, then an unusual jointly administered British-French condominium, 480 miles north-northeast of New Caledonia. "Santo is the real tropics," the young traveler informed his fiancée, "and vastly different from New Caledonia. Hotter, wetter, lusher. And a more interesting place I have never seen." Although the island was only about forty miles wide, its extensive and largely undisturbed rain forest would feed his curiosity; that ants had never before been collected there for scientific study would gratify his opportunism, since every data point he added to his notebook would be new.

Luganville, where he arrived, was the only town on the island, Wilson told Irene, "nothing more than a scattered string of stores & houses, mostly in old U.S. army Quonset huts." He had arranged through a friend on New Caledonia to stay with a prosperous French planter family, one of the wealthiest of some two hundred families on Espiritu Santo who grew coconuts for their copra (the dried meat, from which coconut oil is pressed).

The Ratards—Aubert and Suzanne and their two teenage sons—welcomed Wilson to their substantial plantation. It was "something out of a storybook," he wrote, conveniently located near virgin rain forest. "There are 70 natives here in a village next to Ratard's house; they were brought from the Banks Islands, and though they are very well taken care of and of course completely free, the whole setup is like Tara out of *Gone with the Wind*."

A Georgia plantation was an understandable association for a Southerner to make, but the book the Ratards had been celebrated in was a more recent fiction, James Michener's 1947 novel *Tales of the South Pacific,* the basis for the hit 1949 Rodgers and Hammerstein musical *South Pacific.* Over dinner one evening, Ratard identified himself to Wilson as the model for Emile, the French planter in *South Pacific* who falls in love with the American navy nurse Nellie Forbush, and recalled Michener's visits as a young navy officer during wartime. Michener himself, on a later return, would remember Madame Ratard's roast chicken and good French wine and "a betel-chewing [Tonkinese] woman with a profane vocabulary [who] struck my fancy and became the character Bloody Mary in my novel." Ratard, he added gratefully, "got me started as a writer."

Ratard told Wilson that the real Bloody Mary still lived on a nearby island, Éfaté. Later, the planter walked Wilson down to the shore of his property to show him Bali-ha'i, "across the Segond Channel," Michener had written, "in real life the island of Malo." Madame Ratard served the best food Wilson had ever enjoyed in his life, he told Irene, "an experience to be remembered, from seafood entrees to delicious tropical fruit ice cream."

The Ratards' sons drove Wilson and a native assistant from the family plantation to the rain forest each morning and picked them up again for dinner and supper. The assistant chopped down trees so that Wilson could study the arboreal ants that lived in the forest canopy—"*real* rainforest," he emphasized, "with giant trees, very dark floor, etc. Gorgeous little parrots and pigeons fly around in the treetops, and in the evening flying foxes, giant fruit-eating bats with wingspan about 3 ft., fly leisurely overhead. Flying foxes are considered a delicacy here and Mme. Ratard has promised to serve one up before I leave." His collecting was going well, he ended the Saturday-evening letter, "new species right & left."

Monday, 10 January, Wilson worked again in the rain forest, "picking up ants and batting off hordes of voracious mosquitoes and flies."

The insects would drive him "crazy," he thought, if he had to spend more than a week working there. His first bout of tropical dysentery slowed him down that day; he "quickly drowned it in sulfaguanidine" and it was nearly over by evening. He was learning a little Melanesian Pidgin, the reduced regional language shared between natives and Europeans in an island world where hundreds of local languages were spoken, to mutual incomprehension. Pidgin for "piano," he'd been told, perhaps facetiously, was "big box he got white teeth you hit he sing-sing." The funniest he'd heard was the Pidgin for violin: "Little box you scratch him on belly he cry."

However crazy the mosquitoes and flies might make him, Wilson retained enough wit to generalize from the ant varieties he found on Espiritu Santo to a larger understanding. The species he collected, he wrote later, were "Melanesian, as expected, Solomon Islands most likely, hence ultimately Asian. I made a general observation on the ecology of these insects that would find a place in my later synthesis of island evolution." He had noticed the relative scarcity of ant species on Espiritu Santo; it was, he realized, "just too distant and geologically young to have received many immigrants." As a result of this reduced competition, some species had greatly expanded their niches, densely populating "a wide range of local environments and nest sites." This phenomenon he would later name "environmental release." It was, he realized, "an important early step in the proliferation of biodiversity."

Then a real tropical fever laid him low—"clobbered" him, he told Irene, with a sore throat and a temperature of 103 degrees. The fever broke on Wednesday, though he was "still a little wobbly" as he drafted the midafternoon letter. He doubted if he'd get any more fieldwork done, since the weekly flight to Nouméa left early Friday morning. "I'm discovering that ailments of the sort which give you minor distress in the States can knock you out in the tropics. While I was uncomfortably sick two things kept running through my mind," he concluded gallantly—"You and the music of 'Swan Lake.' The thought of you made me feel quite happy and secure." He enclosed an orchid he'd picked for her. Inevitably, it arrived in ruins, but the thought counted.

Wilson still hadn't tasted flying fox, the big fruit-eating bat of the Santo rain forest. Locals had collected three of the animals on Sunday, and that evening, Wilson had helped clean them of parasites. Madame Ratard had put the local delicacy on the menu for Monday's dinner, which he'd missed because of his fever. He tried warmed-over flying fox Wednesday night, but he was no South Pacific gourmet: "It was terribly rank and gamy," he told Irene, "tasting just about what you'd expect from bat meat, and I couldn't get down more than 2–3 bites."

On Friday morning, 14 January 1955, feeling fit again, Wilson caught the return flight to Nouméa. He found a month of Irene's "beautifully written" letters waiting for him, including a portrait photograph of her that he pledged to keep "sitting in front of me at all stops" and a collection of *Li'l Abner* comic strips, a favorite of his with its Southern setting and perhaps also its barefoot young hero, a hillbilly Candide.

And then, after a six-hour flight from Nouméa on Sunday, he arrived midafternoon in Sydney, Australia, "settled in a comfortable hotel in the downtown section near the Sydney Harbor Bridge," and "at suppertime . . . went out and had my first Chinese dinner in 6 weeks." (Chinese dinners, savory and inexpensive, had been a favorite of the dating couple in Boston.) The Sydney streets were filled with couples strolling hand in hand on that warm January night of antipodean summer, "and believe me," he told Irene, "I ached for you."

The first leg of Wilson's South Pacific expedition had been a great success. "I mailed off the New Caledonian collection to Harvard this morning," he'd written Irene on 5 January, "a real treasure house for future research, over 300 separate collections, all with field notes. I estimate I collected a minimum of 82 native species here, of which nearly half are new to science."

He felt deep gratitude for the opportunity Harvard had given him: "Seldom does a biologist, and especially a young one, get to travel like this to all of the significant rainforest areas. I'm stacking in loads of valuable experience that will temper all of my future research and writing." He elaborated on that theme later: "Ideas in science emerge most readily when some part of the world is studied for its own sake. They follow from thorough, well-organized knowledge of all that is known or can be imagined of real entities and processes within that fragment of existence."

To Irene from Sydney he reported successfully connecting with his Aussie counterparts, both professional and amateur. She had worried that his single-minded intensity could be socially isolating. "I have made friends faster during my Australian visit than ever before in my life," he reassured her, "partly I think because I seriously wanted to make friends here, as many as possible." He was, he added, "carrying through my threat to get along better with people." With its touch of defensive irony, his "threat" sounds more like a pledge his fiancée had drawn from him.

Wilson's primary purpose for visiting Australia was to find an ant of the greatest rarity, the dawn, dinosaur, or living-fossil ant, *Nothomyrmecia macrops*,* previously collected only once, in the winter of 1931–32, and never studied in the field. To that end, he would join a trio of collectors in the South Australian town of Esperance on the last week in January, to search the dense mallee-eucalyptus scrub eastward, where the ant had been found two decades earlier.

Ants (and wasps and bees) are believed to have evolved from parasitic wasps, which attack and paralyze hidden prey and lay their eggs in or near their preys' hideouts as food for wasp offspring. (A more evolved modern form are the familiar mud-dauber wasps, which make the Pan's-pipes-like mud tubes that show up attached to the exterior walls of buildings. Mud daubers paralyze prey, pack it into tubes to feed their larvae when they hatch, and cap each arrangement with an egg.) It's easy to see how this behavior might evolve into the more complicated, specialized, and successful group behavior of the

* Although it takes a little effort to learn Latin binomial names, they identify species more precisely than common names, which (as here) can vary informally. Emphasis in pronunciation usually falls on the third-to-last syllable, thus "no-tho-mir-MECK-ee-uh," the name of the genus, and "MAK-rops," the species name.

social insects, which lay eggs in subterranean channels or tubes of wax or paper and feed the larvae there, but with one or more queens doing all the egg laying, and specialized workers caring for queens and young while others defend the colony or collect food.

Nothomyrmecia, a genus with only one known species, was thought likely to be less specialized than more evolved types; its workers collect prey individually and return it to the colony individually rather than passing it back or carrying it back cooperatively: the dawn ant had not yet evolved trail laying. It was, Wilson wrote Irene, "the most primitive of all ants, and its rediscovery would be an important scientific event. We hope to be lucky enough to study it alive in the field." If they were so lucky, they'd be far ahead of their most recent predecessor, Wilson's Harvard mentor Bill Brown, who had searched the area around the town of Esperance and nearby Mount Merivale four years previously without luck. Wilson's renewed hunt was a good example of his collection strategy: planning broad searches to amass a large number of new species, but going for the rare and spectacular as well when the opportunity arose.

After only a day and another night in Sydney, wasting no time, the eager young entomologist flew on to Adelaide, South Australia's coastal capital, and from there to Kalgoorlie in the Australian Far West, a total distance of 1,747 miles (2,812 kilometers). Beyond Adelaide, "the vast . . . Nullarbor Plain over which the transcontinental railroad runs" impressed him, he wrote Irene, "very flat and very brown." Kalgoorlie reminded him of a small town in Kansas or Oklahoma, pleasant and welcoming. From there he would travel by air and by train almost due south to Esperance, on the south coast, stopping halfway down, in Norseman, "a sunbaked little town," to look for *Nothomyrmecia* alone.

But despite Wilson's best efforts, and those of the three Australian colleagues who joined him in Esperance when he arrived there on 22 January 1955, they found no dawn ants in any of the places where they looked across the next weeks. In the Thomas River basin, their primary search area, they collected both by day and by night, excavated and swept, but *Nothomyrmecia* eluded them. Another twenty-two years would pass before an Australian team would find and study the rare species, a thousand kilometers (about six hundred miles) east of the Thomas River basin, in October 1977. It turned out to be active during the cool antipodean summer season rather than the hot winter season, when Wilson and his colleagues had sought it.

Returning from Esperance, Wilson serenaded his fiancée with an ode to Australia that celebrated his delight in wilderness:

> What a country! Hundreds and hundreds of miles of rough little roads and byways without a habitation along them or even an advertising sign now and then, just tens of thousands of square miles of eucalypt forest and sandplain, from the coast inland to the central desert. Leaving Norseman and heading east, you don't encounter another human (except for a rare passing car) for 186 miles, when you reach Balladonia, with three or four whites and a few aborigines. Then you can turn south, and go for 150 miles or thereabouts to the coast at Thomas River, passing only one lonely sheep station, whose owner is half aborigine and spends much time bush-roaming and tracking dingos. And this is not desert—most of it is covered with fairly tall timber. No wonder Australia is appealing for more immigrants. It's great to see at least one country that isn't overcrowded.

All in all, Australia had been "much more profitable experience-wise than I anticipated," Wilson wrote Irene on 21 February from Perth, as he prepared to fly north, to the largest island in the Pacific. "But now I'm rarin' to get to New Guinea, one of the last and greatest strongholds of stone-age man and the primeval forest and my premier destination on this trip." He might have been traveling backward in time: he'd found no dawn ant in rapidly modernizing Australia, and though he'd looked all over for a boomerang, he couldn't buy one anywhere.

Australia had administered much of long, bird-shaped New Guinea since 1949 as the Territory of Papua and New Guinea; a last-minute seat on a Qantas flight carried Wilson to its capital, Port

Moresby, early in March 1955. He was struck immediately by the less Westernized conditions, he wrote Irene, even in the capital; nearly all the men had "ceremonial scars on their bodies, they stare at you in the street, and (ahem) the women are hardly dressed at all in some cases." He told his fiancée he expected his fieldwork to be "the most exciting of my life." The mountains were "real mountains, many towering over 10,000 ft., the forests are vast and primeval and largely unexplored on foot, and the people are almost a nation unto themselves, over a million strong, with many different customs and languages. . . . Just seen from Port Moresby, New Guinea has a quality of bigness, primitiveness, and natural savagery that I've never encountered before. It's a world by itself, and you could spend a lifetime exploring it. . . . I think I'm going to like the place."

New Guinea has supported human settlement for at least forty-two thousand years. Its interior is so broken up by mountain ranges, dividing and isolating its many indigenous populations, that its people speak no fewer than 851 languages—more than the peoples of any other landmass on Earth. No larger political entity had connected its tribes until the Australians arrived under United Nations auspices at the end of World War II; its Hobbesian anarchy had manifested itself in a homicide rate of as much as 1,000 per 100,000 population, the highest in the world. (The U.S. homicide rate in 2017 was 5.3 per 100,000; the British, 1.2 per 100,000.) The sexes lived sharply divided lives, the men hunting with bows and arrows but seldom sharing the small game they shot (the largest native mammal on the island was a possum-sized marsupial, the cuscus). Women were thus forced to live almost entirely on roasted sweet potatoes, large spiders, and grubs. As a consequence, in the Eastern Highlands around the turn of the twentieth century, women had begun eating their dead in secret feasts, a practice still common at the time of Wilson's visit. Such gourmet cannibalism, recycling brain tissue contaminated with an unidentified organism, resulted in the appearance of a disease that the Fore people of the Eastern Highlands called *kuru,* meaning shivering or shaking. A parallel outbreak in England in the 1990s among cattle that were fed recycled meat-and-bone meal containing contaminated nervous tissue would take the name "mad cow disease." Wilson collected well to the northwest of the *kuru* outbreak areas during his time in New Guinea, and makes no mention of the disease in his records, but the mysterious epidemic and the cannibalism that spread it indicate just how undeveloped the country was.

From his first camp—inland from Port Moresby, on the Brown River—Wilson wrote Irene on 9 March that in the New Guinea rain forest he had reached his "final destination, the place I've dreamed of. . . . My camp is pitched right in the heart of it—giant trees festooned with leaves are on all sides, their green canopy shutting out nearly all light. . . . Every level teems with life—the racket made by parrots and other birds, frogs, and insects, beats on the ears day and night. Vines, tree trunks, and rotting logs almost crawl with an endless array of insect species, including ants, which keep me hopping around enthusiastically." Many of the ant species he encountered "live[d] in silk bags hung from the sides of trees or leaves, and there are army ants here with legions of hundreds of thousands of workers." Rain forest, he concluded, was "the forest primeval . . . ageless and the cradle of all life." Too much life, it seemed; "the excessive heat & humidity, especially at midday," thickened the air, air truculent with "endless, enormous, aggressive, consuming hordes of mosquitoes that are after you every minute of the day. The mosquitoes are the most fatiguing part of daily life here; if they would vanish then it would be a pleasant place to live."

As for living, Wilson wrote, he had four assistants, called "boys" in Pidgin, "a cook-boy, a laundry-boy, a driver and shoot boy, and a general flunky, each being paid 22–33¢ a day plus rations. I've never had it so good—my only problem is finding things to keep them busy, and then struggling with Pidgin English to get my orders across. They've already built me a bed, table, stockage, and fireplace, enough to keep us going 4 months instead of the actual 4 days we'll be out here." Puckishly, he promised Irene to try "not to get spoiled with all this service, so I'll be sufferable when we're married, and I promise never to get up in the morning and shout at you 'Kaikai, you bring 'em, 'e come.' ('Bring my breakfast.')" Back in Port Moresby, on 11 March, he reported with satisfaction that his Brown River camp had offered "just about the richest collecting for a similar period I have ever had; I calculate I collected over 50 species during 3½ days, many of them undoubtedly new."

After two busy weeks in the Port Moresby area, collecting, organizing, and annotating specimens and shipping them back to Harvard, Wilson flew to Lae—due north 188 miles, on the opposite coast—planning to explore inland from there. A week at a lumber camp allowed him to collect and study arboreal ants from the rainforest heights. But a more exciting prospect had opened: a young Australian agricultural officer, Bob Curtis, had invited him to join a monthlong expedition into the Finisterre Mountains, "an expedition of the old tradition, into one of the most rugged, least known parts of the world." Curtis was staging the expedition out of Finschhafen, a harbor town sixty miles east of Lae with a European population of twelve, and on 30 March, Wilson moved there to join him. Curtis told Wilson they would have about twenty carriers, "a new batch picked up at each village . . . just the sort of thing you see depicted as jungle-traveling popularly in cartoons & the cornier Hollywood movies, Curtis & I at the head, Curtis with gun and me with butterfly net, and a string of carriers behind."

The two young men, Curtis twenty-one and Wilson twenty-six years old, left Finschhafen by truck on Monday, 4 April 1955, crossed the Mape River by canoe, and picked up another truck, which delivered them to the coastal village of Heldsbach. At noon, they departed from Heldsbach in grand style, "with 3 regular native assistants, 1 police boy, and 47 carriers." They passed the Lutheran mission at Sattelberg at four in the afternoon and paused to inspect a scattering of signs the Australian army had left, "marking skirmish points in the area where they drove the Japanese into the mountains in 1944." New Guineans had helped and guided the Aussies in their fight against the Japanese during World War II, one reason Australia was supporting New Guinea's development postwar.

It was only fifty miles as the crow flies from Sattelberg to Mount Salawaket, in the Saruwaged Range, Wilson's goal, at 13,500 feet one of the highest peaks in Papua New Guinea, but Curtis planned a month-long expedition to get there, advancing on foot on jungle trails, fording crashing river gorges, "through the middle of the peninsula across the well populated Hube country," Wilson wrote Irene, instructing her to study her map, "to the Cromwell mountains, over the Cromwells at 9,000 ft. to the northern slopes of the central ranges at Iloko, finally from Iloko, the last village, up to the top of Salawaket," and then returning home. Wilson would be the first zoologist to visit the mountain, and he would also "be acting as a genuine geographical explorer—Curtis & I are taking altimeters and will check names of mountains, villages etc. To add to the romance of the thing, the Saruwageds are supposed to be freezing cold, and so many natives have died trying to cross (*I'll* be safe, sweet), the range is known locally as 'mountain belong dead man'—Dead Man's Mts."

He apologized to his fiancée for the length of time he'd be unable to post his daily letter; he would write her daily anyway, the letters forming a diary, and mail all those together when he returned to civilization. With this apology allaying his conflicted feelings, he let loose his enthusiasm: "At the moment," he concluded, "I am very fit, morale is high, and I'm raring to go. The patrol is apt to be the adventure of a lifetime."

It came to less than he had hoped it would, Wilson wrote Irene when he returned to Finschhafen on 23 April, because Curtis had been called back from patrol only three weeks in, before they could climb Salawaket. Still, it had been a grand adventure. By 7 April, well up the peninsula, they had climbed the first mountain in the Hube region, passing over the crest at five thousand feet amid a forest of small trees hung heavily with wet, dripping moss. "The forest floor was very dark and wet and contained few ants or other insects," Wilson told his fiancée. "But leeches were abundant, and the carriers were soon bleeding profusely about the feet from leech bites. When we halted for a rest near the village of Homohang (4500′) an hour later, it was a gory sight, with blotches and rivulets of blood and live leeches on just about everybody's feet, but the natives didn't seem to mind." Leeches are predatory worms closely related to earthworms. To facilitate their feeding, they produce a peptide, hirudin, that prevents blood from clotting; bleeding from leech bites was a familiar experience for tribespeople who worked barefoot and nearly naked in wet mountain forests.

Beyond Homohang was already far enough inland to make European visitors a rarity. In a village in the next valley, Curtis drew awed gasps when he removed an upper dental plate for the adults and children who crowded around them; Wilson scattered the crowd by putting his glasses on backward and pointing and walking in that direction. "One little boy broke out crying," he told Irene. "Few of these people have ever seen eyeglasses, and the children at least really believe I can see through them out of the back of my head."

Four days later, in another village, where Curtis was giving instruction in coffee growing, a group of men flushed a five-foot elapid snake from under the meeting house. Elapids are relatives of the mamba and the cobra, extremely poisonous. Wilson, comfortable handling snakes from his Boy Scout days, dealt expertly with the menace. "There was much excitement and confusion," he wrote, describing the scene, "and when I pinned the snake behind the head with a stick and picked it up, a great hue and cry went up—the natives were deathly afraid of these snakes and with good reason. The whole population crowded around the house looking in while I killed it and put it in preserving fluid."

From there, well inland, they turned south, climbing upward through rain forest, leading a line of thirty-four carriers, passing forty-seven hundred feet in mist and heavy rain. "The place has a miserable climate, due to altitude," Wilson noted on 13 April, "and is probably always under mist or rain during the wet season. The temperatures hang in the 50's & 60's, and believe me, when you are acclimatized to the tropics, this is *cold*."

By Irene's birthday, 19 April, the expedition had turned around and was walking out. "Happy Birthday!!" Wilson wrote his fiancée, adding an eloquent word-portrait of the landscape as a birthday present:

> As I walked back to the village at dusk, the Bulum valley presented a strange and beautiful spectacle. Unbroken forest extended for a thousand feet down to the river and across for nearly ten miles to the Rawlinson range; all was bathed in an aquamarine haze, so that it was like looking down into a deep ocean pool. Cockatoos circled in lazy flight over the treetops like brilliant white fish following the bottom currents. The only sounds I could hear were their faint cries and the distant roar of the river. For me it was a different and very satisfying view of the Forest Primeval.

Four days later, the patrol disbanded in Finschhafen, where Wilson found a letter from Bill Brown reporting that the collections he'd already sent to Harvard were a great success, and "even if you quit this minute, the collection so far is well worth all the cash ladled out."

A greater benefit of the months of travel and work, Wilson wrote Irene, was "a brand new theory" he had conceived during the inland patrol "dealing with the distribution and evolution of animals in the rainforest." He would work on that theory for another year before publishing a major paper in 1958, one of his first contributions to theoretical biology.

In May, shortly before he left New Guinea, Wilson had a second chance to scale the summit ridge of the Saruwaged Range. This time he made it, "the first white man to go to the top of the central portion of the range," he wrote Irene, "and the first to walk up the upper Bumbok valley." The climb was doubly exciting, he elaborated, because he was the first recorded climber of the mountain group of whom he was aware, and because he went through unexplored territory to get there:

> There are very few places left in the world where you can do that. It was a *rugged* trip—the toughest I've ever made. Five days walk to the top (12,000 ft.), the last four by little-used native hunting trails over unimaginably rugged country. . . . Even my guides, recruited from the last native village in the Bumbok, got lost a couple of times. The trip was crowded with events right out of a storybook: uncertain weather making the final climb doubt-

ful but staying clear on the fifth day, natives hunting kangaroos on the summit with bows and arrows, etc. You'll see some of it yourself if the roll of Kodachromes exposed on the climb turn out. These shots are real *National Geographic* stuff—the first color photos from inside the Saruwaged.

He was still tired, but contented. The climb, he concluded, "really [made] my New Guinea tour complete."

Nothing of his work across the next four months would compare to New Guinea, Wilson told Irene in a 4 June letter. "The major part of the trip is now over. There are just 3 other short phases—Queensland, Ceylon, & Europe—all of which are like cleaning-up operations and should move quickly, I hope." He was feeling lonely on that Saturday night in Lae, wondering what his fiancée was doing halfway around the world in Boston, "and how happy we'd be if together there now, even just on weekday nights and Sunday afternoons—strolling, working in the lab, planning a weekend away. . . . Our first year will be exciting in so many ways." He listed some of those ways, anticipation blunting loneliness, "a completely new world for both of us."

Leaving New Guinea, Wilson flew to Sydney, where in mid-June he sailed on the Italian liner *Toscana* around the south end of Australia to Ceylon. (A British colony at the time, Ceylon became a republic, Sri Lanka, in 1972.) He collected on that green teardrop that hangs off the southern tip of India for three weeks, finding 130 species before boarding the liner *Australia* at the end of July for a five-week voyage to Genoa, Italy. "This trip is fairly racing to the end," he wrote Irene before he sailed, adding parenthetically, "(I've started counting *days* now)."

Much of Wilson's time aboard ship he spent reading in the humanities, a deliberate course of self-improvement. In Perth, he'd acquired a dozen Penguin paperbacks with titles ranging from *Psychiatry Today* to H. D. F. Kitto's *The Greeks*. Psychology fascinated him; it was, he wrote Irene, "perhaps the most powerful single subject I've ever encountered outside of my cherished evolutionary theory. It's fascinating just by itself, but [also] offers a key to the understanding of a multitude of other, difficult subjects." Most "modern writers," he observed astutely, "who are attempting a critical survey of their special subject use modern psychological theory as one of their principal tools," citing as examples the art historian Herbert Read, the anthropologist Margaret Mead, and his older Society of Fellows colleague Crane Brinton, a historian of ideas.

Wilson thought he fit Brinton's definition of "a mild 'anti-intellectual,'" although Brinton's concept of this category of thinking made it "very honorable. The 'anti-intellectual' approach is really anti-pure-rationalist," Wilson clarified, but he thought "anti" was an unfortunate label, "which I'll have to discuss with him back at Harvard." The "anti-intellectual," as Crane defined him, Wilson explained, "recognizes the tremendous influence of the unconscious mind and conditioned reflex on human thinking, and the great inertia of ethical behavior. He doesn't believe man can be changed in a generation by rationalist schemes, and therefore tends to be, among other things, anti-communist. He is a conservative among intellectuals, and by some to the left might even be called reactionary." Other than Brinton's "anti" label, Wilson found the characterization compatible, although it hardly encompassed his considerable and rapidly enlarging intellectual range. Being labeled reactionary was a traumatic burden he had not yet been forced to bear. The time would come.

Through August 1955, Wilson worked at museums in Genoa, Paris, and London, comparing his South Pacific discoveries with specimens collected there as long as 160 years ago. From London, on 30 August, he summarized for Irene the significance for them as well as for others of the ten-month odyssey she had understandably called a "nightmare":

> The 1,000 species I have collected personally on this trip, plus the several exchanges engineered with other museums, are adding tremendously to our collection at Harvard and strengthening it as the best of its kind in the world. I'll always be able to draw on it for unlimited research material, even if and when we leave Har-

vard. Actually, on this trip I've laid a firm foundation for years of important research, both for myself and others. The benefits will just begin after I've come home. Wait 'til you see the collection—you'll be amazed at the stuff your fiancé has gathered round the world.

Through the first days of September, to the couple's increasing frustration, airline flights from London to the United States were sold out, filled with homecoming tourists, but sometime after 7 September, Wilson flew to New York and caught the train up to Boston. "Finally," he wrote later, "clad in khaki and heavy boots, crew-cut, twenty pounds underweight, and tinted faint yellow from the antimalarial drug quinacrine, I fell into Renee's arms." ("Renee," pronounced "RE-knee," was Wilson's pet name for his fiancée.)

On 30 September 1955, Edward Osborne Wilson and Irene Kelley were married in the colorful old Roman Catholic church of Saint Cecilia's Parish, in Boston's Back Bay, built in 1894. The wedding was simple. Irene's parents attended, as did Wilson's mother and a few friends, but no bridesmaids or best man. "We both just wanted to get married and get on," Wilson told me. By then, Harvard had offered him an assistant professorship. The Wilsons bought a house in Lexington—west of Cambridge, in the direction of Walden Pond—and began a marriage that has lasted a lifetime.

Once, he had been happy to be entirely alone. In nature he had found "a sanctuary and a realm of boundless adventure; the fewer the people in it, the better." Wilderness for him then was "a dream of privacy, safety, control, and freedom." Those had been the years of his difficult childhood and young adulthood, when he grew into science and mastered the practice of his abundant field.

2

Lost Worlds

In a deep south burdened with history, Edward Osborne Wilson, Jr., born in Birmingham, Alabama, on 10 June 1929, traced his roots back through a government accountant, a river pilot and Confederate veteran, a New England furniture maker, a marine engineer, and generations of farmers. All his immediate ancestors had been Georgians and Alabamians, almost all Southern Baptists. None were college graduates, not even his father, the accountant, who had learned his trade in the United States Army in the years after World War I. Young Ed—"Sonny" to his family—was an original.

A knot of Wilson's earliest memories dates from his seventh year. In the summer of 1936, he was allowed to spend his days wandering alone on a beach on a bell-shaped point of land at the mouth of Perdido Bay, on the Florida-Alabama state line, fifty miles southeast of Mobile. He recalls uncommon events from that summer in his 1994 memoir, *Naturalist:* a young man walking by with a revolver who told him he was out to shoot stingrays; porpoises playing close offshore; a shadowy ray much larger than the common kind, which he tried to catch with a big hook baited with pinfish left dangling off the dock at night but always found stripped bare the next morning. But two crucial encounters he freights with emotions more intense than simple curiosity. One was an alien visitation. The other was an accident.

"I stand in the shallows off Paradise Beach," Wilson begins his description of the alien encounter, "staring down at a huge jellyfish in water so still and clear that its every detail is revealed as though it were trapped in glass. The creature is astonishing. It existed outside my previous imagination." He describes it: "Its opalescent pink bell is divided by thin red lines that radiate from center to circular edge. A wall of tentacles falls from the rim to surround and partially veil a feeding tube and other organs, which fold in and out like the fabric of a drawn curtain." He wants to examine it but is afraid to wade deeper. He doesn't know its name. Sixty years on, recalling the encounter in his memoir, he veers off to pacify it with multiple names—"a sea nettle . . . *Chrysaora quinquecirrha*, a scyphozoan, a medusa"—and then classifies and deconstructs it with his science.

He had no such remedies at hand when he was seven, on Paradise Beach. He had only his unlettered curiosity and a powerful feeling of hazard. "It came into my world abruptly," he recalls the sea nettle, "from I knew not where, radiating what I cannot put into words except—*alien purpose and dark happenings in the kingdom of deep water.*" Just as Herman Melville, in *Moby-Dick*, found in whiteness a terrifying blankness, evoking an "instinct of the knowledge of the demonism in the world," so, across a long life, did Ed Wilson's encounter with a sea nettle in the summer of 1936 still embody "all the mystery and tense malignity of the sea."

What disturbed Wilson so much that he remembered it indelibly sixty years later? "There was trouble at home in this season of fantasy," a subsequent paragraph begins. "My parents were ending their marriage that year." Wilson recalls hearing them yelling back and forth across the apartment where they lived. To give themselves space to work out their future, they had decided to board their seven-year-old only son—precocious, to be sure—alone for the summer with strangers.

A fishing accident had then compounded this strain. Young Ed had hooked a pinfish, a small, silvery Gulf fish with twelve sharp spines on its back, "a favorite among young anglers," according to a parks-department bulletin, "because they are fun to catch." The boy had jerked his fishing rod as soon as the fish struck his minnow hook: "It flew out of the water and into my face. One of its spines pierced the pupil of my right eye."

The pain was excruciating. Remarkably, the boy didn't want to lose the day and suffered the pain so he could continue fishing. "Later," Wilson writes, "the host family, if they understood the problem at all (I can't remember), did not take me in for medical treatment." The pain subsided across the next several days, but at the end of the summer, back at home in Pensacola, the boy's parents found his eye clouding with a traumatic cataract.

He was given surgery to remove the lens, but this brought further trauma, the "terrifying nineteenth-century ordeal" of ether-drip anesthesia through a gauze mask, forced on children in those days unexplained while orderlies held them down. Losing consciousness under such asphyxiating ether-boarding, young Ed dreamed he was alone in a large auditorium, screaming, tied to a chair. "Today," Wilson writes, "I suffer from just one phobia: being trapped in a closed space with my arms immobilized and my face covered with an obstruction."

A summer alone, a sea monster, a partial blinding, terrifying anesthesia, a family breaking up: as if all that confusion and trauma weren't enough, the boy's parents then looked for a safe place to park him again as they separated and pursued divorce. They found a school Wilson recalls as "a carefully planned nightmare engineered for the betterment of the untutored and undisciplined," the Gulf Coast Military Academy, located outside Gulfport, Mississippi, 130 miles distant from house and home. "Though in many of its aspects this visible world seems formed in love," Melville concludes his peroration on nature's blank opacity, "the invisible spheres were formed in fright."

For Wilson, looking back across his life, these experiences as a seven-year-old illustrate "how a naturalist is created." If he hadn't known before, his six months at the military academy taught him the futility of complaint and the honor of mastery. "Hands-on experience at the critical time," he writes, "not systematic knowledge, is what counts in the making of a naturalist. Better to be an untutored savage for a while, not to know the names or anatomical detail. Better to spend long stretches of time just searching and dreaming." He had searched and dreamed; he had even looked a little way into Melville's blank "invisible spheres." That encounter, formed in fright and expectation, had encouraged him to determine to illuminate them. But, untutored savage that he was, he didn't yet know how.

Wilson's mother was awarded custody when his parents divorced. She could not yet support him, however, so his gambling, chain-smoking, alcoholic father took charge. His father's work—auditing rural electrification programs—was itinerant; the boy and his father lived in boarding houses until Ed Senior remarried. After that, Wilson recalls moving among "Pensacola, Mobile, Orlando, Atlanta, the District of Columbia, Evergreen (Alabama), back to Mobile, back to Pensacola, and finally Brewton and Decatur in Alabama, with intervening summer sojourns in Boy Scout camps and homes of friends in Alabama, Florida, Virginia, and Maryland." Across nine years, from fourth grade to high-school graduation, he attended fourteen different public schools.

"A nomadic existence made Nature my companion of choice," Wilson writes, "because the outdoors was the one part of my world I perceived to hold rock steady. Animals and plants I could count on; human relationships were more difficult." After mustering out of the Gulf Coast Military Academy in the summer of 1937, when he was eight, the boy was again farmed out, this time to a family friend, a surrogate grandmother he called Mother Raub. He lived in Pensacola with her and her husband, a retired carpenter, throughout the school year. He skipped a grade that year, which made him "the runt of my class," and found himself fighting off bullies and growing shy and introverted. He recalls being "just as happy to be entirely alone. I turned then with growing concentration to Nature.

Wilderness adventure was popular in an era when radio and motion pictures were just emerging, and when photography had begun to illuminate the world of magazines and books. Wilson thrilled to the exploits of the zoo collector Frank Buck in RKO Pictures' *Bring 'Em Back Alive* ("I don't kill 'em—I bring 'em back alive!"). He read Arthur Conan Doyle's 1912 thriller *The Lost World*, in which a group of explorers discover an elevated plateau deep in the Amazon jungle where living dinosaurs and ape-men roam. The leader of the group, a brilliant, cranky zoologist named Professor George Edward Challenger, tutored a small savage in exploration. "It was my business," Challenger tells the novel's journalist narrator, speaking of an earlier Amazon exploration, "to visit this little-known back-country and to examine its fauna, which furnished me with the materials for several chapters for that great and monumental work upon zoology which will be my life's justification."

In 1938, young Ed's father and his new stepmother, Pearl—"a country lady from King's Mountain, North Carolina," Wilson calls her—collected him and moved to Washington, D.C., where Ed Senior was taking up a two-year assignment at the Rural Electrification Administration. Now nine years old, the boy had become a dedicated reader of *National Geographic* magazine, the gold standard of wildlife and wilderness reportage. Its articles exposed him to the world of insects, "big, metallic-colored beetles and garish butterflies, mostly from the tropics." He was memorably inspired by a report in a 1934 issue, "Stalking Ants, Savage and Civilized," by the director of the nearby National Zoo, W. M. Mann, luridly subtitled, "A Naturalist Braves Bites and Stings in Many Lands to Learn the Story of an Insect Whose Ways Often Parallel Those of Man."

Mann described the Formicidae in all their astounding variety, from "ants as savage and ruthless as the ancient Huns or Mongols," to ants that "make their own gardens and grow their own special food. . . . Ants that keep 'cows'; others that gather and store honey in barrels made from living nest-mates." A suite of full-color paintings illustrating the story included a battle scene captioned "red amazons with ice-tong jaws deal death in a kidnapping raid."

Mann's article, Wilson writes, "led me to search for these insects." He dreamed of prospecting in faraway places. That would come; in the meantime, the one faraway place available to a nine-year-old beckoned only blocks away from his family's basement apartment at Fourteenth and Fairmont. Washington's Rock Creek Park, authorized in 1890 as the nation's third national park, opened three square miles of wooded landscape in the heart of the capital. In a 1918 report, its architects describe "large stretches of forest, [a] river valley, dark ravines, steep and rolling hills, and occasional meadow lands"; to a small boy come north from the Alabama semi-wilderness, it was first of all an insect cornucopia.

There young Ed ventured on expeditions. "Insects were everywhere present in great abundance," Wilson recalls. "Rock Creek Park became Uganda and Sumatra writ small, and the collection of insects I began to accumulate at home a simulacrum of the national museum." The National Museum of Natural History was within easy range as well, a nickel streetcar ride away, and the National Zoo that Mann directed was almost next door, at the head of the park. Both were free and open seven days a week.

"I was catching these big, accidentally introduced Chinese green mantises," Wilson told me. "They were all over the place. Still are. I wrote stories about them in the sixth grade. My parents got a report card about me and my stories which said, 'Ed is doing well. He writes well and he knows a lot about insects. If he puts those two things together, he might do something special.' Yes. That can't have hurt."

That year, young Ed acquired a new best friend and fellow explorer, Ellis MacLeod, a boy a year older who lived down the street and went to the same school. The two boys attended sixth grade together. MacLeod was a butterfly collector; when he heard that Ed also collected butterflies, he sought him out to tell him he thought he had seen a red admiral in the park. A striking black-winged butterfly with white spots near the tips and red median and marginal bands, *Vanessa atalanta* occurs throughout North America and down into Mexico and Guatemala as well as worldwide, adults living on tree sap, fermenting fruit, and bird droppings. The two boys searched the park together, but the red admiral had moved on.

Then, in a further imprinting, young Ed exposed a mystery as compelling as the astonishing opalescent-pink sea nettle had been.

"About this time," Wilson recalls, "I also became fascinated with ants." Exploring the park one day with MacLeod, clambering down a steep, wooded hillside, he pulled away the bark of a rotting tree stump "and discovered a seething mass of citronella ants underneath." Citronella ants live underground and in decaying wood; the worker ants young Ed exposed "were short, fat, brilliant yellow, and emitted a strong lemony odor." Years later, Wilson would report on the source and function of the odor, a compound of citronellal and citral emitted from glands in the ants' mandibles that serves as an alarm substance and a weapon when the colony is disturbed. "That day the little army quickly thinned and vanished into the dark interior of the stump heartwood. But it left a vivid and lasting impression on me. What netherworld had I briefly glimpsed? What strange events were happening deep in the soil?" These questions echo the boy's earlier curiosity about the ocean depths from which the sea nettle had emerged.

Ed and Ellis spent hours wandering the halls of the National Museum, Wilson writes, "absorbed by the unending variety of plants and animals on display there, pulling out trays of butterflies and other insects, lost in dreams of distant jungles and savannas." But their museum explorations opened a larger view as well, Wilson remembers:

> A new vision of scientific professionalism took form. I knew that behind closed doors along the circling balcony, their privacy protected by uniformed guards, labored the curators, shamans of my new world. I never met one of these important personages; perhaps a few passed me unrecognized in the exhibition halls. But just the awareness of their existence . . . fixed in me the conception of science as a desirable life goal. I could not imagine any activity more elevating than to acquire their kind of knowledge, to be a steward of animals and plants, and to put the expertise to public service.

"My future was set," Wilson concludes. "Ellis and I agreed we were going to be entomologists when we grew up." (Ellis MacLeod went on to a lifetime career studying green lacewings and teaching entomology at the University of Illinois. He died in 1997.) Both boys were focused at that time on butterflies. When young Ed returned with his family to Mobile in 1941, he enlarged his butterfly collection with the rich new fauna of the Gulf Coast: "snout butterflies," Wilson reminisces, "Gulf fritillaries, Brazilian skippers, great purple hairstreaks, and several magnificent swallowtails—giant, zebra, and spicebush." With a sweep net fashioned from a cone of cheesecloth, a hooped coat hanger, and a length of broom handle, he could see and capture butterflies well enough. But ants continued to push their way into his awareness. In the summer of 1942, when he was just thirteen, a discovery he made in a vacant lot next door to his house introduced him to the rewards of systematic work.

The Wilsons' house in Mobile had been in the family since the 1840s, when Wilson's great-great-grandfather Henry J. Hawkins, a naval engineer from Providence, Rhode Island, had built it on Charleston Street, only five blocks from the docks along the Mobile River where the river opens into Mobile Bay. By the time Wilson lived on Charleston Street as a small boy, in the 1930s, the big old house, in decline along with the family fortunes, had become a place of retreat for the Wilson kin.

"Big enough to hold three families," Wilson writes, "it was filled to just that capacity in the early years of the Great Depression." In 1932, Ed Senior had moved in with his wife and his namesake son, then three years old. "We joined Uncle Henry and his small family . . . [and] did tolerably well in small apartments staked out on the first floor." Uncle Herbert, a disabled World War I veteran who worked as a security guard on the Mobile docks and who read the *Alley Oop* comic strip to the boy every day, lived in a first-floor bedroom. Upstairs, the second floor was the refuge of young Ed's grandmother and two maiden great-aunts, off-limits to a small boy. It was to this now dilapidated house that Ed Senior had returned his family from Washington in 1941.

At the end of the school year, Ellis MacLeod came down from Washington to spend the summer. By 1942, young Ed was collecting black-widow spiders as well as butterflies. He kept the territorial and poisonous female spiders in Mason jars with aerated lids on a back-porch table, one to a jar, and no one seemed to mind. "We visited my favorite haunts," Wilson remembered of the summer with his friend, "shared again our old fantasies, and renewed our intention to become entomologists."

Spurred perhaps by their renewed commitments, young Ed decided to do a full-scale survey after MacLeod returned home. "Now I'm thirteen years old," Wilson told me. "It was natural at that age that I would want to find out all of the things I could find. There was a vacant lot next door to our house with all sorts of junk in it, weeds, and bushes. My fantasy was to find and map the nest of every kind of ant in that lot. I thought that would be a good thing to do."

That fall, he did it. "I examined every cubic foot of that grubby abandoned space, using a sweep net and crawling on my hands and knees. I covered that place back and forth." He didn't yet know the names of the species he found—he wasn't that far advanced from savagery—but he mapped them. He conducted a self-invented version of what biology calls an ATBI—an all-taxa biodiversity inventory—in this case, of ant species in his next-door vacant lot.

He found exactly four species of ants. Later, he learned their scientific names. The first was *Odontomachus haematodus*, the trap-jaw ant, a big black carnivorous South American native with long, snapping jaws and a painful sting. Young Ed traced the colony to a pile of dirt and roof shingles dumped under a fig tree; the soldiers' vicious stings drove him away from a close inspection.

The second was *Pheidole floridana,* a small fat yellow ant of a dominant genus, *Pheidole,* with more than six hundred species identified in the New World alone. The little colony in the vacant lot was nesting under an empty whiskey bottle; the boy would continue to observe it through the winter, hoping to collect its queen, and find its workers sunning its larvae and pupae under its whiskey-bottle solarium in the cold. Eventually, the queen emerged. Young Ed collected the colony then, moved it inside, housed it in an artificial nest—sand poured between two separated glass plates—and managed to keep it alive for a month while he studied it. In 2003, when he was seventy-four, Wilson completed and published a 794-page large-format scientific catalogue, *Pheidole in the New World,* work that rounded out his lifetime of observations of a spectacularly successful genus.

The third ant species young Ed encountered was *Linepithema humile,* the black imported fire ant, a small, black, invasive pest familiar throughout the South. Multiple queens in vast colonies—one supercolony extends today along the California coast from San Diego north beyond San Francisco to Ukiah, a distance of 615 miles—make them difficult to eradicate. "In warm weather," Wilson remembers, "long columns of this species foraged out into the lot."

Wilson calls the fourth and final species he found in the Mobile vacant lot that autumn of 1942 "the find of a lifetime—or at least of a boyhood." These red ants teemed on a foot-high mound of dirt the colony had excavated in the far corner of the lot. The boy knew immediately that, despite all his collecting, he had never seen such ants before. By their sharp, burning sting he recognized them as fire ants, but they were different from the black fire ants with which he was familiar: red, mound-building, with small heads and toothed jaws, most of all abundant as the black fire ants had never been. In fact, as he would learn later, the strange red ant was rapidly decimating the black ants' colonies wherever the newcomers encountered them.

"I knew I had a new kind of ant," Wilson told me, "but what does a thirteen-year-old kid know about reporting or researching?" Within a year, the curious mound-building ant of young Ed's vacant-lot survey began to win notice as the pestiferous scourge it would become. He had spotted the red imported fire ant, an invasive species that had entered the United States by freighter from Argentina, where it had spread initially from its native Mato Grosso region of Brazil. Much of the cargo shipped into Mobile came from Argentina and Uruguay. Ed Senior, as a teenage seaman, had sailed from Mobile to Montevideo, Uruguay's capital, and back. When young Ed first encountered the red imported fire ant in the vacant lot, no one had yet reported its invasion; he was among the first to recognize its appearance, though it had probably arrived sometime in the 1930s.

It was slowly spreading out from its point of arrival. "Established colonies grew in size swiftly," Wilson would write later. "They began to produce and disperse new queens, hence new colonies, within one or two years." Across the next decade, the relentless march continued, five miles outward per year, an enlarging shock wave colonizing Mobile and invading beyond:

> Soon what had begun in Mobile became a national problem, then global. The imported ant spread to the Carolinas, then Texas and California. It put ashore on Hawaii and made beachhead in Australia, New Zealand, and China. . . . It filled lawns, roadsides, and farmlands, reaching up to fifty mound nests per acre, each teeming with as many as two hundred thousand poised-to-attack workers. On farms in the surrounding counties [of Alabama] the ants consumed seedlings of radish, alfalfa and other money crops. They rendered pastures used for cattle difficult to attend. Their workers even managed to forage into rural houses.

Wilson's life would intersect with the red imported fire ant more significantly in the years to come. Now he was deep into study and work—"in effect a child workaholic," he calls himself in *Naturalist*. He delivered more than four hundred newspapers every morning by bicycle around central Mobile, hard labor for which he was paid thirteen dollars a week. That sum, the equivalent of about two hundred dollars today, was enough to pay for bike parts, candy, movie tickets, and the equipage of the new extracurricular passion that filled his days to the neglect of schooling: climbing the ladder of self-directed learning from rank to rank in the Boy Scouts of America. "All that I had become by the age of twelve, all the biases and preconceptions I had acquired, all the dreams I had garnered and savored, fitted me like a finely milled ball into the socket of its machine when I discovered this wonderful organization. The Boy Scouts of America seemed invented just for me."

Scouting is in decline in the United States today, but it was in its prime at the outset of World War II, with a membership of 1.6 million, boys only. (Between 1942 and 1945, Boy Scouts collected for the war effort 210,000 tons of scrap metal, 590,000 tons of waste paper, and enough milkweed floss to stuff nearly two million lifejackets.) It took seriously its purpose of preparing boys to become honorable men through a military-flavored mix of badges and uniforms, outdoor activities, and extensive rankings, Tenderfoot to Eagle, built on educational and vocational activities and tests. Young Ed was bored with school; he found the subjects he liked most, "outdoor life and natural history," in his *Boy Scout Manual:* "camping, hiking, swimming, hygiene, semaphore signaling, first aid, mapmaking, and, above all, zoology and botany, page after page of animals and plants wonderfully well illustrated, explaining where to find them, how to identify them. The public schools and church had offered nothing like this. The Boy Scouts legitimated Nature as the center of my life." He made Eagle in three years, with forty-six merit badges—more than double the twenty-one required—adorning his merit-badge sash. "The Scout program," he concludes, "was my equivalent of the Bronx High School of Science," the elite New York public high school where many distinguished scientists first found inspiration.

Scouting extended young Ed's social life. It also gave him a first taste of teaching. In 1943, as a fourteen-year-old deep into his snake period, and with older Scouts, of military age, gone off to war, he was invited to serve as the nature counselor at the Mobile-area Boy Scout camp, Camp Pushmataha, named for an important Choctaw chief who died in Washington in 1824 while trying to protect his people from removal. "Snake" Wilson properly awed his fellow Scouts with his snake-handling skills, although an overconfident encounter with a pygmy rattlesnake, which sank its fangs into the tip of his index finger, led to a rush to the doctor for wound suction and a week at home, convalescing painfully on a couch. The camp director thereafter banned poisonous snakes from Snake Wilson's cages.

Back again with Mother Raub that year, encouraged by her steadfast, uncritical faith, young Ed found religion and almost as quickly lost it. He heard the call in a hymn that a tenor visiting his school sang *a cappella* at a recital one evening, a hymn he associated significantly with "the loss of a father" when his own father was increasingly lost to alcoholism. The tenor sang, somberly, "Were you there when they crucified my Lord? . . . When they nailed him to the cross?" Wilson remembers that he was "deeply moved," and wept. That was when he felt "emotion as though from the loss of a father," but it seemed a loss "retrievable by redemption through the mystic union with Christ—that is, if you believed, if you really believed; and I did so really believe, and it was time for me to be baptized."

Meeting the pastor of Mother Raub's church to schedule the baptism was a shock: the Reverend Wallace Rogers was wearing a sport shirt in his church office and smoking a cigar. Mother Raub had exacted a lifetime promise from young Ed never to smoke, drink, or gamble—his father's vices, Wilson notes ruefully—and here was her religious leader breaking the first of those urgent rules. It seemed to the boy a desecration, and Mother Raub's silence about it, then and later, a kind of hypocrisy.

Nor did full-immersion baptism "in a large tank of chest-deep water in the choir loft at the front of the church" supply the "mystic union" the boy craved. To be bent over backward and dipped beneath the water felt unmystically physical, "like putting on swimming trunks and jumping off the tower at the Pensacola Bay bathhouse," and "somehow common." Seeking transcendence, the boy had found only awkwardness and embarrassment, a chilly dip into a common tank of water. As he dried off and returned among the congregation, disenchanted with the baptism's ignition failure, "something small somewhere cracked. I had been holding an exquisite, perfectly spherical jewel in my hand, and now, turning it over in a certain light, I discovered a ruinous fracture."

Recalling the experience in *Naturalist,* Wilson associates it with one of the deepest questions skeptics ask about science as well as religion: "Was the whole world completely physical, after all?" At least in recollection, the baptism marked for Wilson the beginning of a commitment to science "as a means of explaining the physical world, which increasingly seemed to me to be the complete world." Science in its uncompromising materiality would become for him "the new light and the way." In the fullness of time, he would even seek to locate religion itself within its compass.

Across the next two years, young Ed began to leave childhood behind. Snakes continued to occupy him, but he learned them now at the level of detail his peers devoted to memorizing the features of their favorite automobiles. Wilson's father had moved the family once again, this time to the small town of Brewton, Alabama, sixty miles north of Pensacola, on the Alabama side of the Florida Panhandle. At fifteen, young Ed befriended Mr. Perry, an Englishman in his sixties who operated a goldfish hatchery west of town, at the edge of a swamp. Ed interspersed long, comfortable talks with the older man with wading the swamp to collect and study many of the region's forty species of snakes. Then another move, in the late spring of 1945, to Decatur, Alabama—340 miles north of Mobile, on the Tennessee River—ended that preoccupation.

In any case, Ed was focused now on saving for college: "paperboy," Wilson recalls, "lunch-counter attendant and short-order cook at a downtown drugstore, stock clerk at a five-and-ten department store, and . . . office boy in a nearby steel manufacturing plant." He did not enjoy the drudgery: "It persuaded me to strive thereafter to my limit in order to go any distance, master any subject, take any risk to become a professional scientist and thereby avoid having to do such dull and dispiriting labor ever, ever again."

By now, Ed knew he wanted to be a field biologist. He understood he needed to choose an organism he could become expert in. As a sixteen-year-old high-school senior, he approached the problem with the same methodical determination he had previously applied to learning about snakes and butterflies and accumulating Boy Scout merit badges. "I looked up every kind of animal I could get information about," he told me, "trying to decide which one to work on. There was a research laboratory in Decatur, on the Tennessee River. They had a collection of bottled fish, every kind of fish you could imagine, and I worked over those, learned most of their names."

Despite this exercise, Ed found himself drawn back to entomology when his search for a little-studied organism led him to a large family of long-legged predatory flies, the Dolichopodidae, that feed on small insects, "a beautiful metallic bottle-green fly," Wilson recalls, common around creeks and ponds in the South and throughout the Americas. His preference for predatory flies (as opposed to dung flies and mosquitoes) seems partly aesthetic: "I liked their clean looks, acrobatics, and insouciant manner." They were, he writes, "little jewels in nature's clockwork." An accident of history directed him away from flies, however. To collect them professionally, he needed professional equipment, including in particular a long, black insect-pin manufactured almost exclusively in Czechoslovakia. In 1945, none were available; Czechoslovakia was just beginning to recover from the disaster of war and had been confined behind what Winston Churchill would soon call the Iron Curtain.

Chance favors the prepared mind. "But then, one day, I was out in our overgrown backyard," Wilson told me, "and I saw a column of army ants going by, coming out of bivouac somewhere. These were diminutive ants, but real army ants, *Neivamyrmex,* on the march like a Roman army column. I followed them, hoping I could see the queen. They climbed the fence, crossed the next backyard, crossed the street, and disappeared into a patch of woodland." He never did see the queen of these ant-colony predators, but he was intrigued by the camp followers at the end of the column—myrmecophiles, ant lovers, insects that follow ants and steal their food. Later, as a college freshman, he would collect *Neivamyrmex* colonies and keep them alive for laboratory study. "One of my first studies," he recalls, "was of tiny beetles that live on the backs of these ants, feeding on their oily secretions."

In all these formative episodes, his partial vision constricted his choices, he writes in *Naturalist,* but in so doing opened others of great potential:

> I was destined to become an entomologist, committed to minute crawling and flying insects, not by any touch of idiosyncratic genius, not by foresight, but by a fortuitous constriction of physiological ability. I had to have one kind of animal if not another,

> because the fire had been lit, and I took what I could get. The attention of my surviving eye turned to the ground. I would thereafter celebrate the little things of the world, the animals that can be picked up between thumb and forefinger and brought close for inspection.

Not love of ants per se, then, but forays into the world converging in one direction: his partial blindness, Rock Creek Park, citronella ants, fire ants, his decision to choose a little-studied species, a shortage of Czechoslovakian insect pins, army ants across the backyard: less eureka and more informed choice, a promising field of study large enough to span a lifetime. And a family, Formicidae, as successful in its social strategies as our own species but evolved in deep alienage—strangeness at one extreme that mirrored the strangeness of family and humanity a lone boy found at the other.

It was all very well to choose an organism to study, but how was Wilson to pay his way through college at all? College had not been a necessity for his forebears; at the end of World War II, it was still considered a luxury. (In 1945, less than 50 percent of the U.S. population had graduated from high school, and only about 5 percent from college.) Worse, his father—on what Wilson calls the "downward spiral of my father's life as a result of alcoholism, which I witnessed with helpless despair"—developed a bleeding ulcer in the winter of 1945, Wilson's senior year in high school, and underwent a nearly fatal bowel resection, which sent him home to an extended convalescence. "I realized I could not depend on him for further support," Wilson writes, "and feared that I might have to postpone college and take work to assist him and Pearl." His mother, remarried by then to a successful businessman, could have helped, and did so later, but he was too reticent to appeal to her, "a proud, closemouthed kid, frankly ignorant in such matters."

In his desperation, Wilson decided to enlist in the army. The GI Bill of Rights, which President Franklin D. Roosevelt had signed into law on 22 June 1944, would benefit him if he enlisted after his seventeenth birthday, in June 1946. Three years in the army, followed by four years of college, would allow him to graduate at twenty-four. Ed Senior and Pearl approved. "So in June 1946," Wilson writes, "I rode a Greyhound bus to the induction center at Fort McClellan near Anniston, Alabama, where I intended to enlist." Had he come of age a year earlier, during the war, he might have been inducted; when the U.S. war began, in December 1941, vision standards had been lowered so much that even volunteers with one blind eye had been accepted for limited service. Instead, Wilson was rejected—"physical standards," he was told, "had tightened with the end of the shooting war"—and what he believed to be his last recourse for college funding was foreclosed.

Outside the administration building, waiting for his bus, angry and bitter, Wilson railed against the unfairness of his situation:

> I vowed that although I had failed here, I would go on, make it through college and succeed some other way, work on the side as needed, live in basements or attics if I had to, keep trying for scholarships, accept whatever help my parents could give, but regardless of what happened, let nothing stop me. In a blaze of adolescent defiance against the fates, I swore I would not only graduate from college but someday become an important scientist.

The University of Alabama, in Tuscaloosa, like most state universities, accepted any Alabama resident who graduated from high school. Wilson applied there next and was admitted. By attending both term and summer sessions, he calculated he could complete a four-year undergraduate program in three years, and did, at a total cost of only two thousand dollars (equivalent to twenty-seven thousand today). He has been a loyal alumnus ever since. His father and mother also helped rescue him from those basements or attics. By the time Wilson matriculated, in September 1946, Ed Senior had recovered from surgery and found work as an accountant at Brookley Air Force Base in Mobile. He paid part of his son's expenses. Wilson's mother, once she learned of his need, contributed as well.

The Alabama campus was crowded with arriving students, many of them recently discharged veterans. The university had dealt with the overflow by acquiring war-surplus buildings, including a spacious military hospital, where Wilson was assigned a private room on the mental ward—literally, a padded cell—from which he would attend classes held largely in Quonset huts and recreation halls, also relics of the recent war.

Unaware of university admission procedures, and with no knowledgeable family members to advise him, Wilson had assumed he would be expected to declare his major immediately and to demonstrate his competence to qualify. He had corresponded in high school with a U.S. Department of Agriculture entomologist named Marion R. Smith, a specialist in insect identification, who had sent Wilson a mimeographed list he had prepared of the ants of Mississippi with a key to their identification. Wilson decided his project at Alabama would be to assemble a corresponding record of the ants of his home state.

"I was going to study the entire state and learn all the ants," he recalls. "I had the ants I had collected so far labeled and displayed in a Schmitt box someone had given me, a tight-fitting wooden box with a foam lining for pinning specimens. So, during my first week as a freshman, I did what I thought was the correct thing: I went directly to the biology department offices and asked to see the chairman of the department. I was ushered into the chairman's office, where I announced, 'I'm up from Mobile. I'm here to show you my project.'"

Whereupon Wilson produced his Schmitt box, opened it, and began explaining what he planned to do. The department chairman, he says, "was a courtly gentleman. He sat very quietly and listened." After a while, the chairman said, "You just wait a minute," picked up his phone, and called someone. When he hung up, he told Wilson, "Come with me." They climbed a flight of stairs to the office of a young professor of botany, Bert Williams. "The chairman had me wait while he talked in low tones with Professor Williams. Then he shook my hand and went back to his office."

Williams discussed Wilson's planned project with him. "Then he said, 'Come with me.' We walked back into the newly opened research area that he was setting up. There was a row of cubicles, with a microscope at each one, a place for books, and so on. He walked over to one and pointed to it. 'That's your cubicle,' he said." The cubicles were for graduate students. The department chairman and the professor of botany had both recognized the newly arrived seventeen-year-old's commitment and potential and welcomed him to the community of scientists. "That's one of the reasons I love the University of Alabama," Wilson concludes. "Isn't that a great story?"

Williams took Wilson under his wing across that freshman year, inviting him along on field trips and arranging a part-time research assistantship for him: using radioactive phosphorus to trace the uptake of that necessary mineral by plants. In Wilson's second year, he joined a small group of war veterans who had chosen to study at Alabama because they wanted to work with a tough, charismatic young assistant professor of biology named Ralph Chermock. Recently arrived from Cornell University, Chermock had brought with him a deep commitment to evolutionary biology and one of the largest collections of butterflies in North America, a collection Chermock and his brother had assembled across their childhood. Wilson's gang of veterans soon began calling themselves the Chermockians.

Chermock was a hard taskmaster, especially with Wilson, who he had concluded was overpraised and overconfident. The sometimes Sisyphean investigations he assigned Wilson, far from alienating the young scientist, left him cherishing his mentor: "The several best teachers of my life," he would write, "including Chermock, have been those who told me that my very best was not yet good enough."

One of the Chermockians, Barry Valentine, a New Yorker whose mother had founded *Seventeen* magazine in 1944, had a car. "Boy, what a difference that makes," Wilson told me:

> On weekends and holidays we struck out across the state, to the farthest corners and back and forth. We pulled the car over to roadsides and clambered down into bay-gum swamps, hiked along muddy stream banks, and worked in and out of remote

> hillside forests. On rainy spring nights we drove along deserted rural back roads, falling silent to listen for choruses of frogs. . . . On other nights we walked the streets of Tuscaloosa, observing and collecting insects attracted to the lights of storefronts and service stations.

Combined among these four veterans and one recruit were budding specialists in fish, amphibians, reptiles, mollusks, beetles, and ants. "All we did all day long was talk insects, then snakes and everything else," Wilson concludes. "That was our gang, and it was glorious."

But fieldwork was not nearly all they did. Chermock had carried with him from Cornell a commitment to biology's Modern Synthesis. Across the previous three decades, that structure of ideas and discoveries had saved evolutionary biology, Wilson writes, from collapsing "into a jumble of natural history observations. . . . The principle of natural selection, the core of the Darwinian theory, [had been] itself in doubt." One of the central works anchoring the Modern Synthesis, the German ornithologist Ernst Mayr's 1942 *Systematics and the Origin of Species,* became "the sacred text of the Chermock circle," Wilson says. In a curious anticipation of Wilson's mid-1950s battle with James Watson, the codiscoverer of the structure of DNA, over the future course of evolutionary biology, Mayr and his predecessors had fought successfully to enlarge the scope of their science. It had been a battle royal.

3

Natural Selection

In 1859, when Charles Darwin published his revolutionary work *On the Origin of Species,* the idea that species changed over time was hardly in doubt. Too many fossils had been exhumed, too many animal varieties compared, for scientists to contest that question, although the theologically inclined still did so. What was in sharp debate in Darwin's time was not the evidence of evolution, but the mechanism.

The mechanism Darwin proposed was natural selection. The adjective "natural" distinguished it from the artificial selection practiced by breeders of pigeons, horses, cattle, and other domestic animals. Darwin's original formulation, on page 5 of the first edition of the *Origin,* bears quoting:

> As many more individuals of each species are born than can possibly survive; and as, consequently, there is a frequently recurring struggle for existence, it follows that any being, if it vary however slightly in any manner profitable to itself, under the complex and sometimes varying conditions of life, will have a better chance of surviving, and thus be naturally selected. From the strong prin-

> ciple of inheritance, any selected variety will tend to propagate its new and modified form.

But what caused the variations that natural selection needed to select from—the sharper claw, the clearer eye, the better camouflage? In artificial selection—in breeding new varieties of domestic animals—the proximate answer was the crossbreeding of existing varieties. How those varieties had come to exist in the first place was a more complicated question. Ultimately, they had been drawn from wild ancestral populations in the course of being domesticated. But Darwin noted that there is much less variation in wild populations than among domestic animals.

To his great frustration, the answer to how natural variation occurred eluded him. When he sent an early copy of the *Origin* to the comparative anatomist Thomas Henry Huxley, his lifelong friend, Huxley responded with praise but spotted the omission. Darwin acknowledged: "You have most cleverly hit on one point which has greatly troubled me; if, as I must think, external conditions produce little direct effect, what the devil determines each particular variation? What makes a tuft of feathers come on a Cock's head; or moss on a moss-rose?—I shall much like to talk over this with you."

Others noticed the omission as well and criticized Darwin for it. Huxley, by then Darwin's staunch defender, spoke up for his fellow scientist's right to ignore the question of how species emerged in the first place and focus instead on "the totally distinct problem of the modification and perpetuation of organic beings when they have already come into existence." Huxley invoked the greatest of previous scientists to defend the point:

> Would it be a fair objection to urge, respecting the sublime discoveries of a Newton, or a Kepler . . . to say to them—"After all that you have told us as to how the planets revolve, and how they are maintained in their orbits, you cannot tell us what is the cause of the origin of the sun, moon, and stars. So what is the use of

> what you have done?" Yet these objections would not be one whit more preposterous than the objections which have been made to the *Origin of Species*. Mr. Darwin, then, had a perfect right to limit his inquiry as he pleased, and the only question for us—the inquiry being so limited—is to ascertain whether the method of his inquiry is sound or unsound.

Given the irony of omitting discussion of the origin of variation from a book grandly titled *On the Origin of Species*, Darwin could hardly leave the question unanswered. The alternatives his critics espoused were religious—species were God's thoughts, "the mental operations of the Creator," claimed the Harvard paleontologist Louis Agassiz. Or they were vitalist—living beings were organized wholes because they contained a nonmaterial vital force, an *elan vital*, the French philosopher Henri Bergson called it, that guided their evolution. Darwin looked instead to environmental challenges to provide the variation that natural selection required, borrowing an interpretation championed in the eighteenth century by the French naturalist Jean-Baptiste Lamarck. A few pages farther along in the *Origin*, Darwin offers an early guess to what causes variation: "I am strongly inclined to suspect that the most frequent cause of variability may be attributed to the male and female reproductive elements having been affected prior to the act of conception." Or, in a science historian's more pointed paraphrase, "Changes in the conditions of life triggered the reproductive organs to malfunction and to thereby produce variation."

Not only the reproductive organs, Darwin held, but also the entire body might be affected by such conditions. "The direct action [on the body] of changed conditions," he asserted in a later book, was enough to cause variation. He gave examples: "the fleeces of sheep in hot countries . . . maize [corn] grown in cold countries . . . inherited gout." Sheep in hot countries might develop lighter fleece; maize in cold countries might develop greater hardiness. In the other direction, inherited gout might be unfitting: "Natural Selection," the English naturalist noted, "almost inevitably causes much Extinction of the less improved forms of life."

In natural selection, Darwin had identified the *second* stage of evolution. But he didn't know, and never fully worked out, what caused the variation upon which natural selection might operate. All his postulated mechanisms assumed that characteristics acquired during an organism's lifetime (or during the lifetime of an ancestor) could be inherited. In fact, as biologists only understood after Darwin's death, inheritance is complete at conception; acquired characteristics aren't directly heritable.

Darwin seems to have concluded initially that a mechanism for producing variation eluded him because it was subtle and fugitive. He postulated that evolution must occur extremely slowly and in minute stages, many of them invisible. Selection via such small, slow changes would require whole geologic eras, and thus at least multiple hundreds of millions of years, to do its work. That estimate grated against the celebrated British physicist William Thomson's contemporary calculation of the age of the Earth based on a simple model of its cooling from a primordial fireball to solid rock.

Thomson, later to be ennobled as Lord Kelvin, had already established his reputation by formulating the second law of thermodynamics when he estimated the age of the Earth as no more than one hundred million years and possibly as little as twenty million years. That range of time was too brief for Darwinian evolution to accomplish its slow changes, which led Thomson to reject Darwin's theory. As it happened, Thomson was off by several billion years (the current estimate of the age of the Earth is 4.54 ± 0.05 billion years), because he was operating from a flawed picture of the Earth's structure. He imagined it to be solid rock, while in fact its deep interior is semiliquid: molten rock that circulates by convection and limits the flow of heat away from the core. Another factor unknown to Thomson was the heat produced by the decay of radioactive elements such as uranium and thorium in the core; radioactivity was not discovered until 1896, late in Thomson's life. That decay accounts for about 50 percent of Earth's heat.

Darwin answered Thomson by appealing to the authority of the Scottish geologist Sir Charles Lyell, who had asserted the Earth's great age in his *Principles of Geology,* a three-volume work completed in 1833 that Darwin had carried with him on the *Beagle* and studied assiduously. He had no such authority to appeal to in the case of natural variation, however. Something caused it, that much was clear, and in writing his two-volume 1868 compendium *The Variation of Animals and Plants Under Domestication,* he decided to introduce a sort of place marker for whatever that something was.

"I have been led," he announced in *Variation,* "or rather forced, to form a view which to a certain extent connects these facts by a tangible method. . . . I am aware that my view is merely a provisional hypothesis or speculation; but until a better one be advanced, it may be serviceable." To account for natural variation, then, Darwin proposed that evolution followed a process he called pangenesis. His model assumed, he wrote, "that the whole organisation [i.e., organism], in the sense of every separate atom or unit, reproduces itself."

He speculated that environmental challenge stimulated the affected cells of an organism's body to produce an infinitesimal particulate material. "Each cell or atom of tissue throws off a little bud," he explained his idea to Huxley. He called these particles "gemmules," borrowing the Latin diminutive of *gemma,* a bud. Gemmules thus generated pass into the bloodstream, "circulate freely throughout the system, and when supplied with proper nutriment multiply by self-division, subsequently becoming developed into cells like those from which they were derived." Further, gemmules "are . . . transmitted from the parents to the offspring," to be developed in the immediate next generation, but might also be "often transmitted in a dormant state during many generations" before they develop. "Lastly," Darwin concludes, "I assume that the gemmules in their dormant state have a mutual affinity for each other, leading to their aggregation either into buds or into the sexual elements [i.e., seed, sperm, egg]. Hence, speaking strictly, it is not the reproductive elements, nor the buds, which generate new organisms, but the cells themselves throughout the body. These assumptions constitute the provisional hypothesis which I have called Pangenesis."

Darwin was not alone in his confusion about variation and reproduction. One scientist, writing in 1903, identified more than thirty different theories of heredity proposed in the second half of the nineteenth century. Inheritance to Darwin, as well as his contemporaries, was blended rather than particulate: rather than the modern view of a set of instructions coded in germ cells, randomly varied and separately inherited, in blended-inheritance theory the whole organism contributed its traits and experiences to its offspring. Darwin's gemmules, little buds that accumulated into sexual or somatic cells that reproduced by dividing, generalized the process across all the many different forms of life, from plants to animals, from microbes to fungus. That Darwin postulated gemmules to be too small to see under a microscope emphasizes the importance of the contemporary development of microscopy and its application to cytology in particular—to the study of cells in all their many forms.

Cytology progressed rapidly in the middle and late decades of Darwin's life (he died on 19 April 1882), driven in part by the development of better microscope lenses. Before that development in the 1830s, chromatic aberration in poorly corrected lenses had introduced confusing halos of color around images, particularly when making observations in sunlight in that time of weak illumination with oil flame or candle. New compound lens systems that corrected for chromatic aberration, along with techniques for staining tissues to increase their contrast, opened a view into microscopic biology beyond what Darwin had been able to compass.

In the last decade of Darwin's life, German cytologists armed with better microscopes and more sophisticated staining techniques were able to move from studying the nuclei of cells to studying their individual components. The German biologist Walther Flemming in an 1880 paper named the small threads within the nuclei of germ cells that easily took staining "chromatin," "stainable material." From that word, another German biologist, Heinrich Wilhelm Waldeyer, coined the term "chromosome" in 1888. It was also Flemming who first described and named the fundamental process of cell division, mitosis, and who demonstrated definitively that the chromatin in the cell nucleus, not the entire nucleus, initiated and controlled the process of dividing. "Cell division involving indirect division of the nucleus has been found in every case studied with the necessary care and the proper methods," he concluded.

Yet, for all this back-to-the-lab tracing of the physical machinery of reproduction, none of these scientists had yet identified what caused the variation that served as the basis for natural selection. Lamarck's eighteenth-century "laws" of evolution argued, first, that an organism's use or disuse of a trait led to the trait's increase or decrease and, second, that such change was heritable. Darwin had cited many examples, as he believed, of the inheritance of acquired characteristics.

Lamarckian inheritance was logically inconsistent. It drew followers in part because it seemed to give purpose to life by making it possible for nature and humanity to work in the direction of improvement, a basically religious perspective. A letter to the British journal *Nature* in 1894 from an Oxford professor of zoology, Edward B. Poulton, concisely identified the logical inconsistency in Lamarck's theory. The French naturalist had postulated an old Earth, one with ample time for extremely gradual change to occur. With that long history in mind, Poulton found a "mutual antagonism" between Lamarck's first two laws. The first law, Poulton pointed out, "assumes that a past history of indefinite duration is powerless to create a bias by which the present can be controlled; while the second assumes that the brief history of the present can readily raise a bias to control the future." In less Victorian language: It doesn't make sense that traits evolved through eons can be changed in the short course of a single lifetime to produce modified traits that then persist for eons. Either traits were mutable or they were not. They couldn't logically be both at once.

By the end of the nineteenth century, biologists were on the verge of rejecting Darwinian evolution entirely—just as the actual cause of natural variation came to light. It had been hiding in plain sight for more than three decades. The work of the Augustinian abbot and Czech scientist Gregor Mendel wasn't obscure, as traditional histories have it. His "Experiments on Plant Hybridization" was published in 1866 in a widely circulated journal; it was even referenced in an article in the ninth edition of the *Encyclopedia Britannica.* Why that breakthrough paper was neglected is a complex story outside the range of this review: partly because Mendel dealt with *varieties* of the peas he studied when the emphasis of evolutionary science at the time was on *species;* partly, as two scholars write, because the paper concerned "the *transmission* of inherited traits . . . and therefore would have seemed to his contemporaries a partial work, a work that could not have contained a theory of inheritance." Darwin died unaware of it.

Mendel's paper, with its revolutionary observations, was in any case rediscovered, by three European botanists working independently in three different countries, and at just the right time—at the beginning of the new century—to turn the study of natural selection away from gemmulean, Lamarckian, and other blind alleys. (Ralph Chermock, Ed Wilson's mentor and sometime scourge at the University of Alabama, was a relative of one of the three rediscoverers, the Austrian agronomist Erich von Tschermak.) Mendel's investigations supplied the missing piece to the puzzle of what generated the variation upon which natural selection worked. His answer was, fundamentally, random mutation, understood today to occur because genes are sometimes damaged or miscopy themselves. Cells repair most such mistakes. Those that remain persist as inheritable dominant or recessive traits that natural selection can then weed out or distribute across a population, depending on their survival value. "Varieties enjoying even a slight competitive advantage," one historian summarizes, "would, over multiple generations and in a process acting like compound interest in banking, come to predominate within a population."

The study of biology needed a generation after Mendel's rediscovery—into the 1930s and 1940s—to work out this fruitful merger of Darwin's and Mendel's insights, with much disagreement and mutual incomprehension along the way. As it had since the beginning, Darwin's discovery of natural selection offended the convictions of those who hoped and believed that the thrust of evolutionary change moved humanity toward improvement. Mendel's rediscovered insights were first used to attempt to discredit Darwinian evolution altogether, with the claim that mutation was sufficient to cause evolutionary change even without natural selection.

Only after that effort failed on the evidence was it possible to merge the two great biological discoveries into one productive whole. The German ornithologist Ernst Mayr, writing about the transition, emphasizes the signal contributions of evolutionary naturalists as compared with experimental biologists. "The particular evolutionary problem in which the naturalists were interested," Mayr notes, "was speciation"—the evolution of new species. "They continued the Darwinian tradition of the Galapagos finches," which famously had evolved a variety of beak adaptations to fill the empty food-gathering niches of those isolated islands far off the coast of Argentina. "And everything they found seemed to confirm Darwin's insistence on gradualism and to refute the saltationism of the Mendelians." (Mayr elsewhere defines "saltation," from the Latin *saltare*, "to leap, to skip, to jump," as belief in "the sudden origin of new species" as opposed to the gradual accumulation of small variations upon which natural selection might act.)

On the other hand, Mayr emphasizes, "there was no doubt in the minds of the experimental biologists that their methods were more objective, more scientific, and therefore superior to the 'speculative' approach of the evolutionary naturalist." Mayr quotes Thomas Hunt Morgan, who founded the famous Fly Room at Columbia University to experiment with mutation using fruit flies. Morgan wrote in 1932 that biology would make progress only by "an appeal to experiment. . . . The application of the same kind of procedure that has long been recognized in the physical sciences [is] the most dependable one in formulating an interpretation of the outer world." A similarly acrimonious competition between laboratory and field biologists would challenge Wilson's early faculty years at Harvard.

Wilson came of age in the midst of this great struggle and achievement, with Ralph Chermock guiding him and his platoon of war veterans as they surveyed the insect fauna of Alabama in the field and around the drugstore and filling-station lights of Tuscaloosa.

When Wilson looked up from studying descriptive natural history as a college student and explored the processes of evolution, his career vision expanded. "I learned to ask: What kind of process creates biodiversity?" he writes. "What other kind scatters species into their current geographic ranges?" He learned that neither kind occurs at random; both were "the products of understandable causes and effects"—which made them accessible to scientific investigation. "I was already totally devoted to making a career in natural history," he recalls of that time, "as an expert on insects. A government entomologist, perhaps, or a park ranger or a teacher. Now I rejoiced. I could also be a real scientist!"

The imported fire ant that Wilson had first encountered at thirteen in the vacant lot next to his home in Mobile was continuing to spread outward from its entry point in Mobile Harbor at the steady rate of five miles a year, like a slow-motion shock wave. It crossed Wilson's path again in 1948, when the outdoors editor at the *Mobile Press Register* quoted him in a series of articles about the increasingly destructive pest. That public notice of Wilson's expertise, now honed by his university studies, led the Alabama Department of Conservation to hire him to conduct a study of the ant and its environmental impact.

"I took leave from the university for the spring term [of 1949]," Wilson writes, "to begin, at the age of nineteen, a four-month stint as entomologist, my first position as a professional scientist." Working with James Eads, a war-veteran colleague who owned a car, he mapped the ants' range, dug up colonies to study their structure, interviewed farmers, and investigated crop damage. "In July we submitted a fifty-three-page analysis to the Department of Conservation office in Montgomery titled 'A Report on the Imported Fire Ant *Solenopsis saevissima* var. *richteri* Forel in Alabama.'" (In 1971, a new investigation led to a name change, to *Solenopsis invicta,* the species name meaning "unconquered," as the imported fire ant was and continues to be.)

Wilson earned his bachelor-of-science degree in biology from the University of Alabama in 1949, his master of science, also in biology, from Alabama the following year. By then, he understood that to be a "real scientist" would require earning a Ph.D., but he knew little as yet about the world beyond the South. "I wasn't very imaginative about leaving Mobile and going to some Northern university," he told me, "someplace like Minnesota, whose name I couldn't even spell. So I said, I'll go north to Tennessee if I can get admitted into the University of Tennessee in Knoxville. There was a professor of entomology there, Arthur Cole, who specialized in ant classification. And I was accepted." Two years previously, Cole had been part of a team studying the fauna on Bikini Atoll. Cole's extensive collections from throughout the United States as well as the Philippines and India gave Wilson a chance, he writes, to hone "my skills in the anatomy and classification of insects," a tedious but necessary competence.

The University of Tennessee was a backwater, however, and Wilson soon became bored. Since his first year of college, he had corresponded with a Harvard Ph.D. student in entomology, William L. Brown. (Wilson remembers himself at the time as "a callow, severely undereducated eighteen-year-old," though by then he was carrying around a talismanic and much-consulted copy of Ernst Mayr's formidable *Systematics and the Origin of Species.*) "Although only seven years my senior," Wilson writes, "Bill was already a leading world authority on ants. At that time there were only about a dozen experts on ants worldwide and he was one of them. . . ." Like Wilson, Brown had been an insect enthusiast since childhood. He grew up in Philadelphia; when his parents took their children to the Jersey Shore in the summertime, they would arm Bill with a bag lunch and drop him off at the New Jersey Pine Barrens so that he could spend the day collecting in that semi-wilderness halfway between Philadelphia and New York. He was a veteran, having served in an air force malaria-survey unit in western China and India during the war. He graduated from Penn State in 1947 and would finish his Ph.D. at Harvard in 1950 and precede Wilson to Australia.

Brown "was . . . one of the warmest and most generous people I have ever known," Wilson remembers. During Wilson's undergraduate years, Brown plied him with advice and encouragement, always in the direction of expanding his horizons. When Wilson proposed to survey the ants of Alabama, Brown urged him to go continent-wide or even global. The upshot was work on a large tribe of ants, the dacetines, first by Wilson alone and later in partnership with Brown.

By 1950, Wilson was aiming for Harvard—his destiny, he believed then, because of the world scale of its ant collection and its "long and deep" tradition of ant study. A botany professor at the University of Tennessee, Aaron J. Sharp, encouraged him to apply and recommended him. "He wrote a couple of people at Harvard," Wilson told me, "and said, 'This kid does not belong here. He belongs at Harvard.'" Brown endorsed him as well. A letter came one day from Brown's faculty adviser, the distinguished paleoentomologist Frank Morton Carpenter, who built and curated the collection of fossil insects at the Harvard Museum of Comparative Zoology, the MCZ. Carpenter invited the young Alabamian to come up and visit.

Wilson made the journey in late June 1950, a marathon seventy-two-hour Greyhound bus ride from Mobile to Boston that seemed to stop at every small town along the way and left him loopy from lack of sleep. Bill Brown and his wife, Doris, welcomed him and let him recover on their Cambridge apartment couch after Brown took him to visit for the first time a room that would become his second home.

The ant room at the MCZ held drawer after meticulous drawer of preserved specimens from across the world, nearly a million in all, each black, brown, red, or yellow ant neatly mounted on a pin with an identifying tag. The pioneer entomologist William Morton Wheeler had founded the MCZ's unsurpassed study collection in 1908, when he arrived at Harvard from New York's American Museum of Natural History to take up a professorship. Wheeler's 1910 book, *Ants: Their Structure, Development and Behavior,* based on lectures he'd given at Columbia University in 1905, was a founding study of ant life. "When I was sixteen," Wilson told an interviewer in 2011, "and decided I wanted to become a myrmecologist, I memorized that book." Now twenty-one-year-old Ed Wilson was standing in the midst of the collection that Wheeler, Carpenter, and their colleagues had gathered from every corner of the world.

During Wilson's visit, several faculty members recommended

he apply to Harvard for graduate study. Back home, he did that, and worked at the University of Tennessee through the fall and winter to assemble a collection for a major study of the origin of the ant caste system, a paper he would publish in *The Quarterly Review of Biology* in 1953. (Castes, he concluded, such as queens and workers, evolved from natural variations in size and anatomy within an ant population.)

His father's health had been in sharp decline for several years. Ed Senior struggled with little success to control his alcoholism. Though he managed to keep his job, he regularly checked himself into rehab, dried out, returned home, and gradually descended into drinking again. Sometime that fall, father appealed to son to give up seeking a Ph.D. and stay home. You have a master's, Ed Senior told Wilson; with that, you can stay in Alabama and get a good job, take care of me, take care of Pearl. Wilson assessed his father's appeal as objectively as he could. On the one hand, his staying home might buy Ed Senior a few years, but it would undercut any possibility of a full scientific career. On the other hand, Wilson argued, "When you total up both our lives, I would give up a career that would potentially benefit us both." So, he concludes, "I said no."

Having failed to pass the burden of personal responsibility to his son, ill and despairing, Ed Senior, early on Monday, 26 March 1951, picked out a target pistol from his extensive gun collection, drove to a remote part of town, sat down beside the road, put his pistol to his temple, and shot himself. He was unconscious and moribund when he was found and died at the hospital later that day. Pearl called Wilson in Tennessee in tears and broke the grim news.

The bus trip from Nashville to Mobile took twenty-four hours. Wilson's father had qualified as a World War I veteran by virtue of a 1920 enlistment, when he was seventeen; by the time Wilson arrived in Mobile, fellow veterans had stepped in to help. Ed Senior, dead at forty-eight, received a military funeral and was buried in Mobile's Magnolia Cemetery, where the Wilson ancestors lay. "I was unhappy that I'd decided not to stay and help," Wilson assesses his reaction, "but it wouldn't have prolonged his life much, and it would have ruined mine." The only message Ed Senior left behind, scrawled on a torn scrap of paper, was "I'm sorry."

Pearl, Wilson's stepmother, a strong woman, soon moved past her grief, found work, and eventually remarried.

Much later, writing his memoirs, Wilson noted that admiration for his father's courage had replaced his earlier feelings of "sorrow and guilt-tinged relief." It would be easy to say of his father, he wrote, "that the greater courage would have been to try again, to pull himself back and struggle toward a normal life. I am reasonably certain, however, that he had considered the matter very carefully and decided otherwise." Ed Senior had always said he was afraid of becoming "like a Bowery bum." Wilson saw his father's suicide as a deliberate act to forestall that outcome: "I think he would have died rather than accept humiliation or disgrace as defined by his lights. In truth, he did just that in the end."

In doing so, he freed Wilson from any obligation to support him in his decline, freed his son to move on. In the international scientific community, which someone once compared to a large village, exceptional scientific talent lacking financial resources often finds quiet recruitment and support from senior scientists. Recruiting promising novices is one important way the tacit knowledge of science—the part that isn't in the textbooks—is passed on to the next generation. Passed on along with that tacit knowledge is often the frame of ideas that the senior scientist originated or endorses.

Wilson in his turn benefited from the support of this informal network. Just when he lost a father, surrogate fathers stepped forward to help him along as he moved north into an exhilarating new world. From his Tennessee botany professor's "This kid . . . belongs at Harvard" to his recruitment there, to the tolerance for his limited basic-science education shown him after he arrived at Harvard as a twenty-two-year-old Ph.D. student in the autumn of 1951, means came to hand. "I was considered a prodigy in field biology and entomology," he confirms, "and was allowed to make up the many gaps in general biology left from my happy days in Alabama." A scholarship and a teaching assistantship paid his way.

He found worthy peers as well, particularly a Harvard senior and then fellow graduate student in entomology named Thomas Eisner—another prodigy, whose background was far from Wilson's. Eisner was Jewish, born in Berlin, an early émigré with his family from Nazi Germany to Barcelona, where the Eisners then found themselves in the middle of the Spanish Civil War—"Stukas dive-bombing the city," Wilson remembers Eisner saying. They immigrated ultimately to Uruguay, where Eisner grew up. "On a grander scale," Wilson assesses, "he had repeated the pattern of my own childhood, having been towed from one locality to another, anxious and insecure, turning to natural history as a solace." He was ten days younger than Wilson, and an equally ardent entomologist, but with a different bent. His father was a chemist who had trained under Fritz Haber, the German chemist who invented a method for synthesizing ammonia from atmospheric nitrogen, the basis of the modern chemical-fertilizer industry as well as Germany's main source of ammonia during World War I for propellants and explosives. Young Eisner would go on after Harvard to a career teaching at Cornell; there, with his colleague Jerrold Meinwald, he would found a new field of biological science, chemical ecology, devoted to studying insect chemical communications and defenses.

The highlight of the two young scientists' time together at Harvard was a twelve-thousand-mile trip they made together collecting ants—"naturalist hobos," Wilson calls them—in the summer of 1952. "Up into Canada," he reminisced to me, "all the way across Canada to the Pacific, down the Pacific Coast through Washington and Oregon and California—we crossed Death Valley on a hot night in July with wet handkerchiefs on our heads—over through the South and back up again to Cambridge, collecting and talking, a great experience." They remained friends for life; Eisner died of Parkinson's-disease complications in 2011 after a distinguished career.

Then came Wilson's election as a Harvard Junior Fellow, in 1953, which he applied first to a field trip to the Caribbean, an early glimpse of environmental devastation. "My intellectual journey gathered momentum in 1953," he writes, "in Cuba's Sierra Trinidad, as I toiled up muddy roads in search of rain forest, past logging trucks on their way to Cienfuegos with the final fragments of the trees." He'd done a lot of exploring in his twenties, he told an interviewer many years later. "Even then, you could see the ravages visited upon the natural world, and I was conscious of extinctions. At that time, I didn't see it as serious enough globally to require immediate action. Maybe I saw the flora and fauna of the world as immortal." Would that they were.

After the Caribbean came the long trek across the South Pacific the following year. In March 1955, picking up his mail in Lae, New Guinea, Wilson heard from Bill Brown that an open instructorship at Harvard was being given to someone else. "Nobody doubts that my credentials were the best around the place," he wrote Irene, "but the selection of the instructors was entirely in the hands of the man in charge of the elementary zoology course, Prof. Griffins, and being a neurophysiologist, he decided to use the job as an opportunity to bring in a younger neurophysiologist to Harvard for a couple of years." Wilson explained to his fiancée his reasoning for gambling on another Harvard opening even though he had an offer from the University of Florida in hand:

> This doesn't mean that we will go to Florida. After much careful thought and some soul-searching I have made up my mind to remove myself as a candidate for the job at Florida, the second year in a row I've done this. I've decided to complete my third year as a Junior Fellow. The decision came from trying to answer the old, familiar question, "shall I take the nice, comfortable job now, or shall I take the one that is lower paying and less secure but offers the most opportunities for training and future advancement?" If I went to Florida, we would have a pretty fair salary right off, comfortable surroundings, and the security of a permanent job. By continuing as a Junior Fellow, on the other hand, we'll be making a thousand dollars less that year (1955–56), the "job" only lasts one year, but I will have the entire year to continue my research with Harvard's facilities and to live in the

> stimulating environment of the Society of Fellows. . . . Prof. Carpenter & Brown have both strongly urged that I stay at Harvard that extra year. There is even some talk of my landing a position at Harvard the following year in the professor ranks, but this is *just talk* at this stage, and should not stir our hopes.

It proved to be more than just talk. After Wilson's return from the South Pacific, the department of biology of Harvard University offered him an assistant professorship, beginning with the 1956 school year. It was what he had gambled on, and he accepted his prize with pleasure.

James Watson, the 1953 codiscoverer with Francis Crick of the double-helix configuration of DNA, had been offered an assistant professorship at Harvard like Wilson's in 1955, had accepted the offer, but had immediately taken a one-year sabbatical to continue work at Cambridge University in England, where the DNA discovery had been made. He and Wilson would thus begin teaching at Harvard at the same time, in the autumn of 1956. But Wilson was a field biologist, Watson a laboratory biologist, and Watson, who believed biology could now best be pursued in the lab, was determined to sweep the Harvard biology department clean of field scientists. He viewed them with derision. "Stamp collectors," he called them.

4

Stamp Collectors and Fast Young Guns

IF JIM WATSON THOUGHT of his fellow assistant professor Ed Wilson as one of his "stamp collectors," Wilson was a great admirer of Watson and Crick's discovery and, initially at least, of Watson himself. "I was among the Harvard graduate students most excited by the early advances of molecular biology," Wilson writes. "Watson was a boy's hero of the natural sciences, the fast young gun who rode into town."

An earlier appreciation, and a book, had prepared the way for Wilson's admiring assessment. In his last year in high school, he had thrilled to the news of atomic energy and the sudden celebrity of J. Robert Oppenheimer, the physically tough but wraith-thin American theoretical physicist who had directed the secret U.S. atomic-bomb laboratory during World War II. "I was especially impressed by a *Life* magazine photograph of him in a porkpie hat," Wilson recalls, "taken as he spoke with [Manhattan Project commander] General Leslie Groves at ground zero following the first nuclear explosion." That first explosion had been a test conducted in the New Mexican desert in July 1945, prior to the atomic bombings in Japan, although the photograph had been taken postwar.

"Oppenheimer was a slight man, as I was a slight boy," the high-school senior had felt. "He was vulnerable in appearance like me, but smilingly at ease in the company of a general; and the two stood there together because the physicist . . . had tamed for human use the most powerful force in nature."

The book that had influenced Wilson, a note in a bottle pitched over from physics to biology, was the Austrian émigré physicist Erwin Schrödinger's slim 1944 volume, *What Is Life?* When Wilson encountered it as a college freshman, he recalls, it "was creating a sensation among biologists. . . . The great scientist argued not only that life was entirely a physical process, but that biology could be explained by the principles of physics and chemistry. Imagine: biology transformed by the same mental effort that split the atom! I fantasized being Schrödinger's student and joining the great enterprise." So much the more threatening, then, that the fast young gun Jim Watson was agitating to clear Harvard of its muddy-booted field biologists and replace them with a posse of experimentalists in white lab coats.

The tension between field and laboratory wasn't new. It had emerged with the emergence of biological science itself early in the twentieth century. By the 1930s, the strain was evident. "Many practitioners of the new experimental biology," one historian of biology writes, "were increasingly turning their backs upon the findings and theories of the older field- and museum-oriented 'naturalists.'" Ernst Mayr would call for "exceptional individuals with a wide range of knowledge and interests," to bridge the gap. Before that happened, the two sides—embodied at Harvard in the late 1950s in champions such as Wilson and Watson—would have to fight it out.

Schrödinger's little book, adapted from lectures the Nobel laureate had delivered in 1943 at Trinity College, Dublin, where he landed in exile from Nazi-occupied Austria, had also influenced Watson and Crick along the way to their great and fundamental discovery; after their announcement in 1953, they would thank the émigré physicist for the trail of clues he had laid. Watson went even further, assigning to Schrödinger's lectures a key role in his decision to turn, ironically, from ornithology—from field biology, that is, which at Harvard he would scorn—to genetics:

> I got hooked on the gene during my third [undergraduate] year at the University of Chicago. Until then, I had planned to be a naturalist and looked forward to a career far removed from the urban bustle of Chicago's South Side, where I grew up. My change of heart was inspired not by an unforgettable teacher but by a little book that appeared in 1944, *What Is Life?*

The core of Schrödinger's speculation was his assertion that what passed from parent to offspring to guide the development of a new individual was not a little copy, a little homunculus, but *information*. Chromosomes, he wrote, "contain in some kind of code-script the entire pattern of the individual's future development and of its functioning in the mature state. Every complete set of chromosomes [found in almost every cell in the body] contains the full code."

Moreover, as Schrödinger emphasized, the "code script" had to be something more substantial than a list of instructions; it had to be something physical, like the notches in a key that line up the tumblers in a lock. A great volume of information had to be transmitted to make anything as complex as a living creature. An encyclopedia of the human genome, for example, would comprise several thousand volumes. How could so much information be stored in anything so small as a chromosome?

Given the size constraints and the load of information, instructions for making and operating new creatures evidently took the form of large molecules, Schrödinger suspected, the arrangement of their atoms presumably forming the necessary code-script. "Large" is a relative term here. If cells are so small they have to be observed under a microscope, even large molecules are significantly smaller. A large molecule within a cell is comparable in size to a cat on a cruise ship. And at cat-and-cruise-ship scale, a grape could represent an atom. Schrödinger speculated that the large, complicated molecule that carried the genetic information—the notched key—

must be what he called an "aperiodic crystal" or "aperiodic solid." A periodic crystal, such as a diamond, builds up from a crystal seed by repeating the same structure in three dimensions again and again. Such a repeating structure, a hum more than a signal, can carry very little information.

"The other way," Schrödinger pointed out, "is that of building up a more and more extended aggregate without the dull device of repetition. That is the case of the more and more complicated organic molecule in which every atom, and every group of atoms, plays an individual role, not entirely equivalent to that of many others. We might quite properly call that an aperiodic crystal or solid and express our hypothesis by saying: We believe a gene—or perhaps the whole chromosome fiber—to be an aperiodic solid." ("Has the man never heard of a polymer?" an organic chemist asked of Schrödinger's clunky formulation. Polymers are chemical compounds organized on repeating structural units; natural polymers include silk, wool, cotton, and . . . deoxyribonucleic acid: DNA.) Francis Crick would write of Schrödinger's formulation: "It was only later that I came to see its limitations—like many physicists, he knew nothing of chemistry—but he certainly made it seem as if great things were just around the corner." Crick, a physicist before he began working in molecular biology, was himself an exception.

Fortunately for Watson and Crick's future achievements, during the 1940s, and even as late as 1952, the candidate material for carrying the genetic code was thought to be a protein—a complicated organic molecule, aperiodic and solid—not nucleic acid. "When I arrived at Harvard as a graduate student in 1951," Wilson recalls, "most outside the biochemical cognoscenti believed the gene to be an intractable assembly of proteins. Its chemical structure and the means by which it directs enzyme assembly would not, we assumed, be deciphered until well into the next century." The Canadian geneticist Oswald Avery and his colleagues at the Rockefeller Institute Hospital in New York had demonstrated in 1944, not long after Schrödinger's Dublin lectures, that DNA was in fact what they called the "transforming principle," at least in the bacterium they were studying that caused bacterial pneumonia. Looking back on that time, one of Wilson's contemporaries, the Nobel laureate geneticist Joshua Lederberg, would recall that the nucleic-acid molecule "was then believed to be a monotonous structure.... The protein-enthusiasm evoked by the successful crystallization of [proteinaceous] enzymes in the 1930s then dominated most biochemists' attention." In any event, Lederberg adds, "Nothing was known of chromosomes or genes in bacteria at that time." It was easy to assume that what happened in bacteria might be different from the far more complex requirements of reproduction in larger organisms.

However misdirected the search for the genetic code, the search was on in the years after World War II. In 1952, a crucial series of experiments, the Hershey-Chase experiments, turned the tide in favor of DNA. Alfred Hershey was a Michigan-born physical chemist who became one of the founding members of the phage group, a circle of scientists who worked together informally, beginning in 1940, studying genetics by using a convenient class of viruses known as bacteriophages—"phages" for short, from the Greek word meaning "to eat"—that infect and destroy bacteria. Besides Hershey, the group included Watson's Indiana University Ph.D. supervisor, Salvador Luria, and the theoretical physicist Max Delbrück. Luria and Delbrück, an Italian and a German, were both refugees from European fascism, technically enemy aliens ineligible for security clearances, and thus free to work on basic science rather than military technology during wartime. Phages had the advantage of easy manipulation and rapid reproduction. They were "an even speedier experimental subject than *Drosophila* [fruit flies]," Watson recalls: "genetic crosses of phages done one day could be analyzed the next." In 1945, the three scientists initiated a summer phage course at the Cold Spring Harbor Laboratory on Long Island's North Shore. Hershey and his research assistant, Martha Chase, made their 1952 discovery there.

Seen under an electron microscope, the type of phage Hershey and Chase used, the T2 bacteriophage, looks like a cross between a moon lander and a hypodermic syringe:

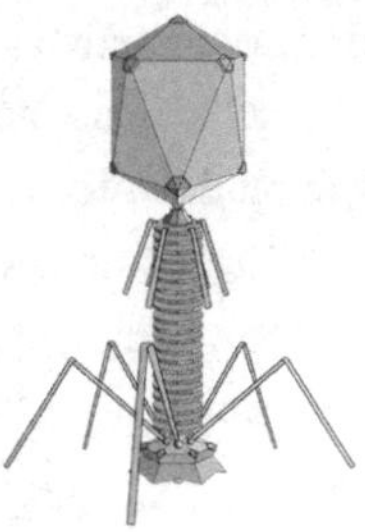

The head of the phage contains its DNA ("like a hat in a hatbox," Watson described it). The legs attach it to a bacterium. The cylinder connecting the two serves as a transfer tube to deliver the phage DNA from the head of the phage to the interior of the bacterium, thus infecting it:

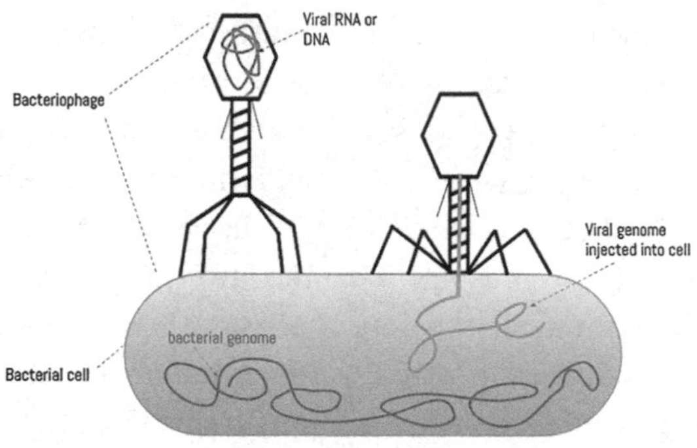

The entire phage assembly except its DNA is made of protein. DNA contains phosphorus, but not sulfur, and protein contains sulfur, but not phosphorus. So Hershey and Chase made two batches of the phage, one grown in radioactive phosphorus 32, the other grown in radioactive sulfur 35. By following the different radioactivities, which served as tracers, they hoped to see which material ended up where. In a series of experiments, they first showed that the phosphorus-labeled DNA left the phage when it infected the bacterium, leaving behind sulfur-tagged protein "ghosts" of the phage membrane—the deflated head and the transfer tube, still attached to the cell surface.

Then, in the final experiment of the series, nicknamed the "blender experiment," the two researchers labeled phage batches with radioactive phosphorus or sulfur, infected bacteria with the labeled phages, then agitated the bacteria in solution in an ordinary kitchen blender to strip off the phage protein "ghosts." (They used a kitchen blender because a more powerful laboratory centrifuge would have broken up the bacteria; the blender removed the protein ghosts but left the bacteria intact.) The bacteria, with their phage ghosts removed but with the phosphorus-tagged phage DNA inside, produced a fresh crop of phage virus, proving they were infected by the DNA alone and not the proteinaceous ghosts. The experiment confirmed what was still being debated: that DNA, not protein, was the sole carrier of genetic information.

Here, then, was Schrödinger's "code-script" and Avery's "transforming principle." Hershey announced the results at an international congress of biochemistry in Paris that summer. "DNA was the hereditary material!" Watson wrote excitedly. ". . . It was more obvious than ever that DNA must be understood at the molecular level if we were to uncover the essence of the gene." Hershey and Chase's result, he adds, was "the talk of the town." Watson realized that the news was sure to draw the formidable Caltech biochemist Linus Pauling into the race.

It did, and early the following year, Pauling published a paper proposing what Watson recognized to be a mistaken model of the structure of DNA. How that mistake led to Watson and Crick's race to work out the actual double-helical conformation of the DNA molecule is a story Watson tells vividly in his colorful and controversial 1968 memoir, *The Double Helix.* Recalling it many years later, he summarizes:

> In *What Is Life?* Schrödinger had suggested that the language of life might be like Morse code, a series of dots and dashes. He wasn't far off. The language of DNA is a linear series of [paired molecules of the nitrogenous bases adenine, thymine, guanine, and cytosine]. And just as transcribing a page out of a book can result in the odd typo, the rare mistake creeps in when all these

A's, T's, G's, and C's are being copied along a chromosome. These errors are the mutations geneticists had talked about for almost fifty years.

The paper that made Watson and Crick famous, and led to their receiving a Nobel Prize along with their colleague Maurice Wilkins in 1962, was published in the journal *Nature* on 25 April 1953. Titled "Molecular Structure of Nucleic Acids: A Structure for Deoxyribose Nucleic Acid," it culminated in a conclusion of wry understatement: "It has not escaped our notice that the specific pairing we have postulated immediately suggests a possible copying mechanism for the genetic material."

Despite the discoverers' instant celebrity, Watson was almost rejected for appointment at Harvard. "He gave a test lecture," his biographer, Victor McElheny, told me. "He was mumbling and whispering. It was not thought to have been a successful communication, so people were really quite nervous about him coming. About taking him at all." Achievement—that dazzling double helix—tipped the scale in his favor.

Then began what Wilson would call the "molecular wars." Ranged on one side were the new and existing Harvard faculty members newly committed to molecular biology, on the other the biologists whom Watson derided as "stamp collectors": the zoologists, botanists, ornithologists, and entomologists who investigated groups of organisms. "The science was now being sliced crosswise," Wilson writes, "according to levels of biological organization, that is, oriented to the molecule, cell, organism, population, and ecosystem respectively. Biology spun through a ninety-degree rotation in its approaches to life."

The small cadre of Harvard biochemists and molecular biologists, as Wilson lists them, included Brooklyn-born George Wald, "soon to receive a Nobel Prize for his work on the biochemical basis of vision"; John Edsall, a New Englander and pioneering protein chemist who was a Harvard classmate and lifelong friend of Robert Oppenheimer; Paul Levine, a population biologist who left that field "and began to promote the new doctrine [of molecular biology] aggressively on his own"; and Matthew Meselson, who moved to Harvard from Caltech in 1960 as an associate professor after important work with his colleague Franklin Stahl showing that the DNA double helix replicates by separating into two single strands. The molecular enthusiasts, one science historian comments, arrived armed with "a confidence in their emerging field that bordered on imperialistic zeal."

They took up residence in the Harvard Biological Laboratories on Divinity Avenue, behind the Museum of Comparative Zoology, the same building where Wilson's office was located, near the entrance on the first floor. Outside his windows, two monumental bronze statues of the great Indian rhinoceros stood sentinel; inside, in the halls, the warring camps collided. The young entomologist found these exchanges distinctly uncomfortable, particularly since Watson persistently snubbed him. "On one occasion," Wilson recalls, "in October 1962, I offered him my hand and said, 'Congratulations, Jim, on the Nobel Prize. It's a wonderful event for the whole department.' He replied, 'Thank you.' End of conversation."

"My self-esteem was fragile then to a degree that now seems beyond reason," Wilson would comment later. Nor was Watson's discourtesy directed at Wilson exclusively. "He treated most of the other twenty-four members of the Department of Biology with a revolutionary's fervent disrespect," Wilson writes. "Few dared call him openly to account." Understandably, Wilson thought Watson's rudeness repellent. "I found him the most unpleasant human being I had ever met. . . . At twenty-eight, he was only a year older. He arrived with a conviction that biology must be transformed into a science directed at molecules and cells and rewritten in the language of physics and chemistry. . . . His bad manners were tolerated because of the greatness of the discovery he had made, and because of its gathering aftermath. In the 1950s and 1960s the molecular revolution had begun to run through biology like a flash flood."

Victor McElheny, Watson's biographer, enlarges the scene to take in Watson's point of view: While Wilson may have felt most keenly his departmental fights with Watson over appointments in ecology and environmental studies, two subjects Watson had little tolerance for, Watson felt he was beating his head against a wall of little fiefdoms, each professor with his own operation organized around whatever species he studied. "But if you're dealing with the transfer of information," McElheny told me, "from the DNA vault out to the factory suburbs where the proteins are formed, that's an extraordinarily complex class of events to try to understand. The amount of data involved really is inherently large, and the instrumental complexity of finding the information and recording it requires not drawers full of neatly labeled specimens but roomfuls of instruments—centrifuges, electrophoresis systems, electron microscopes, all kinds of spectroscopes."

To tackle such volumes of information with such tools, McElheny went on, you have to assemble persons of multiple talents. "So you're driven, not necessarily to big science, but at least to sharing instruments and to all kinds of modern forms of cooperation. You need a larger organization and a looser one, and perhaps a more brutal one as well, because you have to smash a lot of idols to get to where you're going." As often with human conflict, what looked from one side like naked aggression looked from the other side like necessary expansion and reform.

One incident between Wilson and Watson that encapsulates this multilayered conflict was Wilson's attaining tenure—lifetime appointment as a member of the Harvard faculty—ahead of Watson. Promotion at Harvard often depends on pressure from outside the organization itself: Harvard professorships are greatly coveted and well-paid measures of outstanding achievement, charily granted by tenured faculty who often doubt if their younger colleagues at Harvard and elsewhere could possibly measure up.

As it happened, however, Stanford University wanted to build a new program of entomology upon the approaching retirement of its one incumbent professor in the field. Wilson's offer from the California school arrived out of the blue, in a letter from the chairman of the Stanford biology department in 1958, when Wilson was halfway through a five-year appointment at Harvard as an assistant professor.

He and Irene were sorely tempted. Stanford's president, Wallace Sterling, a historian and champion fund-raiser, was building a first-class faculty and campus at that time across what was still a sunny orchard and ranching landscape south of San Francisco, in Palo Alto. Wilson was even more tempted when the dean of the Stanford faculty and Sterling himself personally visited the twenty-nine-year-old in his Biological Laboratories office one day. A proposed salary of seventy-five hundred dollars—the equivalent of about sixty-eight thousand today—enriched the offer. "And Stanford would assist us in buying a house," Wilson recalls, "a policy unheard of at Harvard."

On the verge of accepting, Wilson stopped by his department chairman's office to thank him for all Harvard had done for him. Frank Carpenter, the chairman, encouraged delay. Wait a few weeks, he told Wilson: let's see what Harvard can do. McGeorge Bundy was Harvard's dean at that time, two years away from moving to Washington as John F. Kennedy's national security adviser. Evidently, he had no intention of allowing the university to lose one of its most promising young stars. With uncharacteristic alacrity, Harvard matched the Stanford offer. Wilson decided to stay where he was, despite the wet, cold coastal-Massachusetts climate. To this day, he wonders if he made the right decision; he has always felt more like an Alabamian in the North than a confirmed Harvardian.

When Watson heard the news of Wilson's accession to tenure, he swore his way through the Biological Laboratories' halls, loudly repeating (depending on who is telling the story) either "Shit, shit, shit, shit!" or "Fuck, fuck, fuck, fuck!"

McElheny elaborates on Wilson's story, extending it beyond the clash of personalities it certainly partly was. "Jim didn't just walk up some stairs in the Biological Laboratories cursing when Wilson got tenure ahead of him. He also went to Nathan Pusey, the president of the university at that time, and screamed at him. I don't know the details of that interview, but it was an extremely angry and abusive confrontation. Watson just unloaded the vials of his wrath on Pusey. It's amazing, because this also bespeaks an almost inexpressible vulnerability on Watson's part." Watson erupted, McElheny implies, because Wilson's appointment looked to him like a loss in the battle for the dominance of molecular biology at a time when the field was exploding with new discoveries.

Something more fundamental than boorishness, jealousy, or faculty politics drove these confrontations. They resembled the confrontations between species that drive natural selection itself. Reductionist biology—reducing biology's complexities as much as possible to its fundamental physics and chemistry—had invaded the science's cultural space. Maneuvering for physical space for offices and laboratories, expanding the curriculum, pulling in valuable grant funds, and winning faculty appointments paralleled an invasive species' physical domination of a territory, expropriating resources and challenging the previous occupants.

"The traditionalists at Harvard at first supported the revolution," Wilson confirms. "We agreed that more molecular and cellular biology was needed in the curriculum. . . . The ranks of molecular and cellular biologists swelled rapidly. In one long drive, they secured seven of eight professorial appointments made. No one could doubt that their success was, at least in the abstract, deserved. The problem was that no one knew how to stop them from dominating the Department of Biology to the eventual extinction of other disciplines."

Wilson was committed to enriching biology with a reductionist program. Watson, he writes, "had pulled off his achievement with courage and panache. He and other molecular biologists conveyed to his generation a new faith in the reductionist method of the natural sciences." But Wilson also believed that biology was more than its molecular and biochemical scaffolding. The natural world in all its immense variety was more than chemistry and physics, more than molecules. Relationships within and between species could be informed by the new science; it could hardly fully explain them. Francis Crick, though an extreme reductionist, clarified the distinction in his 1966 book *Of Molecules and Men:* "In examining every biological system," he wrote there, "one can always ask how it works; meaning how, from a knowledge of its parts, one can predict its behavior. Alternatively, one can ask how the system got that way; in other words, how it evolved." Organismic and evolutionary investigation would continue to be fruitful, enhanced by the new molecular perspective. The majority of living species on Earth had not yet even been named, much less studied in depth.

Characteristically, Wilson responded to the challenge of the molecular invasion by networking and communicating, operating now from the greater security of his tenured associate professorship. "In 1960," he writes, "the faculty members of the Department of Biology working on ecology and evolution, being outgunned and outfunded and soon to be outnumbered, decided to form a committee to organize and unify our efforts." One outcome of that decision was an opportunity to fix an appropriate and clarifying new name upon what the molecular biologists were calling, derisively, "classical" biology, and the embattled traditionalists were calling clumsily, for lack of a better name, "macrobiology."

In the spring of 1958, Wilson had coined the term "evolutionary biology" as a course title and entered it into the Harvard course catalogue. Now, waiting in a seminar room for the committee to assemble for its first meeting, he broached the term with another early arrival, his distinguished but notoriously uncommunicative colleague George Gaylord Simpson, one of the founders of the Modern Synthesis. Wilson reconstructs their dialogue:

> "What shall we call our subject?" I ventured.
>
> "I have no idea," he responded.
>
> "What about 'real biology'?" I continued, trying for humor. Silence.
>
> "Whole-organism biology?"
>
> No response. Well, those were bad ideas anyway.
>
> There was a pause, then I added, "What do you think of 'evolutionary biology'?"
>
> "Sounds all right to me," Simpson said, perhaps just to keep me quiet.
>
> Other committee members began to file in, and when all were settled, I seized the opportunity to assert, "George Simpson and

I agree that the right term for the overall subject we represent is 'evolutionary biology.'"

And so it became—almost. Ultimately, the only way to resolve the dispute between the evolutionary biologists and the molecular biologists was to split into two departments, much as religions split when their members can't agree on their fundamental beliefs and are unwilling to compromise. The "department of molecular biology" went one way. Wilson's department became the "department of organismic and evolutionary biology"—a clumsier name than he might have liked, but one that signaled compromise was possible, at least on the evolutionary side of the faith.

If the molecular-biology challenge had threatened Wilson, it had also energized him, encouraging him to move on from his early studies of taxonomy—of finding, preserving, and identifying specimens—to broadly evolutionary investigation. "What we were now calling evolutionary biology," he writes, ". . . required a rigor comparable to that of molecular and cellular biology. We needed quantitative theory and definitive tests of the ideas spun from the theory and vivid connections to real-life phenomena."

Among much other effort, including in particular identifying young potential colleagues who were "as able and ambitious as the best molecular biologists," Wilson invested himself in two projects. One was improving his math skills, the better to work the complexities of quantitative theory. "I finally got around to calculus as a thirty-two-year-old tenured professor at Harvard, where I sat uncomfortably in classes with undergraduate students only a bit more than half my age. A couple of them were students in a course on evolutionary biology I was teaching. I swallowed my pride and learned calculus."

The other project was more fundamental and more promising. Wilson began following up on a goal he had first identified in 1953, when the later Nobel laureate Konrad Lorenz had lectured at Harvard on the new science of ethology, the scientific study of animal behavior. Lorenz had spoken then of graylag geese and jackdaws and stickleback fish. Wilson, in a sudden flash of inspiration, had seen that ethology could also encompass insect behavior—and, in particular, the secrets of their communications. The big animals communicated with sight and sound. Insects, moving underground and in darkness, communicated with chemicals. How did they do that? What chemicals did they use? Where in their bodies did they produce them? Before he could investigate that largely unknown domain, though, he had to finish his Ph.D., explore the South Pacific, begin teaching, and establish himself in his department. Now that work was in hand. Now he could find out how ants talk to one another. Maybe he could even talk with them himself.

5

Speaking Pheromone

DIGGING INTO an ant colony in the field sends workers and queens scattering everywhere. That was no way to observe them, much less to experiment with them. To experiment with ants, Ed Wilson worked out a way to keep them in the laboratory. Since his undergraduate days in Alabama, he had improvised various kinds of tubs and boxes for ant keeping. By the time he moved on at Harvard from collecting and classifying to doing experimental work, in the late 1950s, he had developed a dependable laboratory ant habitat: a multilevel Plexiglas box about three feet long and two feet high. A sliding wall allowed access for cleaning or manipulation. The floor between the levels was made of softwood strips; a curved plastic tube extended through the wall to allow the nest to be watered. A single door in the bottom gallery of this artificial nest opened to a walled foraging area—an outdoor pen, as it were—where the colony could be observed as it displayed its behaviors. The nest interior was brightly lit, light to which the ants had been gradually habituated. Fluorescents illuminated the foraging area but operated on an automatic twelve-hour timer, simulating the cycle of day and night. The entire habitat could be covered with a Plexiglas lid, but Wilson normally left it open for easy observation and handling. A coating around the rim, of a commercial fluoropolymer resin, Fluon, which dried too slippery for the ants to cross, kept them from escaping.

Wilson's preferred study subjects were fire ants, despite their small size, about one-sixteenth of an inch, and their painful stings. They were plentiful and easily acquired. On weeklong excursions, Wilson would drive to South Carolina or Alabama, alone or with a graduate student or two, park on a rural roadside beside a fire-ant mound—they were everywhere now in the South—fetch a garden spade and a bucket from the trunk of his car, deposit a flattened beer can nearby, then dig the spade deep into the mound and scatter a spray of dirt and agitated ants along the roadside. Inevitably, he'd suffer stings, but a few fire-ant stings were a small price to pay for a living colony to transfer to his habitat in the Harvard Biological Laboratories. In the mêlée of the scattered colony, the ant queen's attendants would spot the flattened beer can and hustle her over to hide her underneath. All Wilson had to do then was to turn over the beer can, catch the queen with forceps—she was many times larger than her crumb-sized workers—slip her into a collecting tube, and pop on the lid. He called the method the Beer Can Technique. A few scoops of workers dropped into the bucket along with the queen in her tube, and some paper toweling for shelter, and he had a functioning colony to transplant to the laboratory. If a highway patrolman came along and stopped to find out what he was doing, he told the officer he was working for the government. That was all it took, and he was, at least technically, since his research grants came from the National Science Foundation.

Back at Harvard, Wilson prepared a nest box and moved the colony in. The queen would soon enlarge it, laying eggs at such a rate that it would have reached twenty thousand workers or more within a year had Wilson not kept it culled. He maintained his research colonies at about one hundred individuals, more than enough for his studies, and easier to feed and maintain.

That ants communicate by smell had been known since at least the late nineteenth century. The Swiss neuropsychiatrist Auguste Forel, an amateur myrmecologist whose major study of the ants of Switzerland had won praise from Charles Darwin, summarized the importance of smell in ant life in his 1908 book, *The Senses of Insects:*

> It can therefore be boldly supposed that the antennae [of ants] and their olfaction [i.e., their sense of smell], as much on contact as at a distance, constitute the social sense of ants, the sense which allows them to recognize one another, to tend their larvae and mutually help one another, and also the sense which awakens their greedy appetites, their violent hatred for every being foreign to the colony, the sense which principally guides them . . . in the long and patient travels which they have to undertake, which makes them find their way back, find their plant-lice and all their other means of subsistence.

But what were they smelling? What secretion or secretions could determine in an instant if one ant welcomed another ant into the nest or furiously attacked it? Bees that discover a field of flowers returned to the hive and performed a waggle dance to teach their sisters where to find the new source of food; ants, operating in two dimensions and often in darkness, apparently laid down a scent trail to recruit their kin. And the actions of ants in response to such chemical markers were compulsory, not voluntary.

Organic compounds that are secreted inside the body to regulate some specific cellular activity are called hormones. In 1959, in the journal *Nature,* two European zoologists proposed a new name for the analogous compounds then being investigated that are secreted outside the body for communication between individuals: "pheromones." In their new coinage, the Greek word from which "hormone" is derived, *ormōni,* meaning "to excite," supplied an ending to another Greek word, *pherein,* meaning "to transfer." "Pheromones," the zoologists wrote, "are defined as substances which are secreted to the outside by an individual and received by a second individual of the same species, in which they release a specific reaction, for example, a definite behavior or a developmental process."

"Humans communicate by sight and sounds," Wilson comments,

"allowing the creation of words with arbitrarily chosen meaning. In other words, we communicate by language, the prerequisite for the most rapid possible evolution of social order. Ants, in contrast, use chemical secretions, smelled or tasted, with gene-fixed meanings."

But no one had yet identified those chemical secretions, nor had their source within the ant's body been found. In the summer of 1958, Wilson decided to try to find and identify them. He had already realized that the trail ants lay down to guide their fellow workers to a newly discovered food source must be a pheromone. He had decided as well that the best way to determine which gland or glands produce the trail pheromone was to test extracts of various glands by laying artificial trails with them and observing how his laboratory ants responded. "If I could find this part of the ant Rosetta Stone," he writes, "I could 'speak' to a fire-ant colony and tell them where to go for food."

So he would be looking for an exocrine gland—one that discharges its secretion through a duct to the outside of the body, as human tear and saliva glands do. That limitation pointed him to the fire ant's mouth at one end or stinger at the other, its body's obvious openings to the outside world. He had already noticed that fire ants, laying trails, curled their gasters—the bulbous rearmost segment of their bodies—under them and touched their stingers intermittently to the ground, "much like a moving pen dispensing ink," which directed him to the gaster as the likeliest place to search.

The body of even a small ant is packed with organs and glands. Fire ants are among the smallest, their internal parts diminutive. Wilson had neither the time nor the money that summer to buy and learn to operate a complex micromanipulator system in his laboratory. Instead, he relied on a large magnifying glass mounted on a bench stand, his one, exceptional eye, and what he calls "the finest of all handheld instruments, the needle-tipped Dumont Number 5 forceps, used by jewelers to handle very small precious stones." Even then, the scale at which he was working ruled out deliberate movement. When he tried deliberate dissection, overcompensation scattered ant parts everywhere. He soon learned to take advantage of the hand's normal tremor, its fine, hammerlike response to the pulsing of the blood. "These involuntary movements were just enough to cut through the base of the glands," he recalls, "and ease them into insect-blood saline for further preparation."

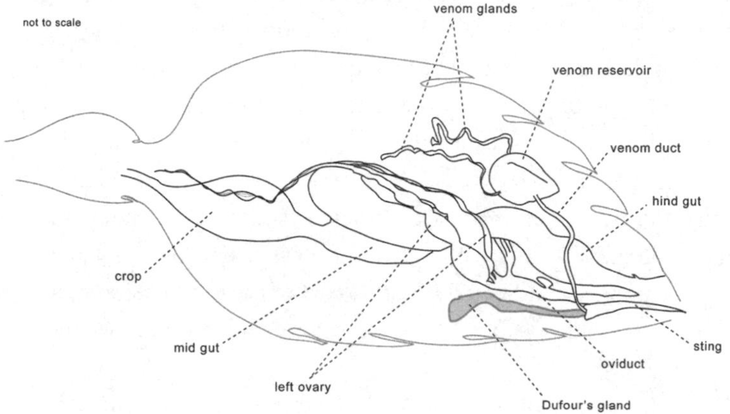

The ant gaster is crowded with glands and other structures.

Since fire ants appeared to mark a trail with their stingers, Wilson looked first at the fire-ant poison gland. "The pheromone might prove to be the venom itself," he theorized. "But when I tested this hypothesis, the result was wholly negative. Venom meant nothing to the hungry fire ants. Nor did, less surprisingly, the contents of glands I tested from other locations throughout the body."

Then he noticed another possible source, a small, sausage-shaped gland that opened into the base of the stinger: Dufour's gland, named for the nineteenth-century French physician and naturalist Léon Jean Marie Dufour, who first described it in 1841. "Barely visible to the naked eye as no more than a tiny white speck," Wilson writes, "it seemed an unlikely candidate for so important a role as the fire ant odor trail." He teased one loose from the opened gaster of a freshly killed ant, washed it in saline, and crushed it against the tip of a sharpened applicator stick. With this unlikely brush, he painted a trail from his fire-ant colony's nest entrance out to the middle of its foraging area. "The result was stunning," he reports. "The ants in the

Biologist Edward O. Wilson (EOW) was born in Alabama in 1929. As a skeptical three-year-old, he was already booted for outdoor adventure.

Ed Wilson Senior, a government auditor, was a drinking and gambling man. Eight-year-old EOW, with an eye lost in a fishing accident, turned to exploring "the little things of the world."

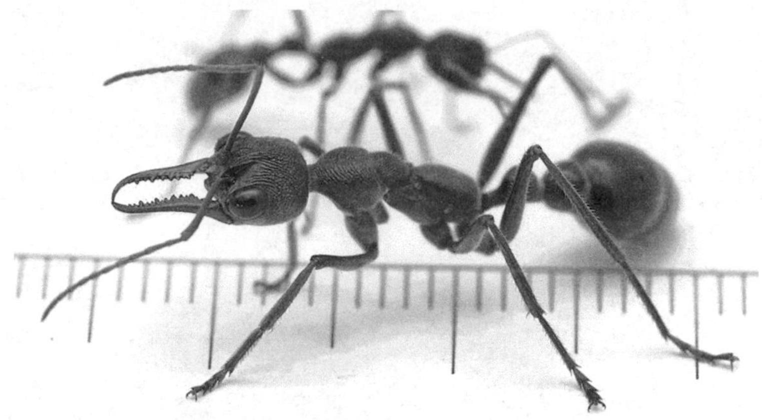

Young EOW thrilled to read of Australia's fierce "bulldog ant" in an old *National Geographic*. He would encounter it as a postdoc on an expedition to the South Pacific.

EOW at thirteen in 1942 was already a dedicated collector, the first in the United States to spot the invasion of the pestilential red imported fire ant.

Scouting, with its focus on nature craft and knowledge, was EOW's equivalent of a high school of science.

EOW's parents divorced when he was seven. Growing up, he lived with his father, but his mother, Inez Freeman Wilson, remained a strong influence.

Awarded a Junior Fellowship at Harvard in 1953, EOW first traveled to Cuba to collect ants. Sugar cane dominated; little remained of wild nature there.

The strength of his marriage to Bostonian Irene Kelley Wilson helped sustain EOW in his academic battles.

In 1954–55, EOW explored the South Pacific for new species of ants. Upcountry from Port Moresby, New Guinea, he found his best collecting.

Stuart Altmann, EOW's first graduate student, introduced him to animal social behavior among the rhesus monkeys Altmann studied in Puerto Rico. For EOW it was "an intellectual turning point."

Fire ants became EOW's favorite experimental subject. Molten aluminum poured into a nest (inverted here) reveals its elaborate structure.

To collect fire ants, EOW shoveled a nest into a nearby pond. The ants quickly formed a living raft from which they could be scooped into a bucket for transport.

Along with lab and field research, EOW taught popular classes in biology at Harvard. The university awarded him tenure in 1958 at the young age of twenty-nine.

EOW's competitor at Harvard, brash molecular biologist James Watson, the codiscoverer of the structure of DNA, pushed to eliminate field biology. EOW led the fight to save it.

Yale's pioneer ecologist G. Evelyn Hutchinson taught EOW's island biogeography coauthor Robert MacArthur. The theory extended far beyond islands to support nature reserves on land as well.

nest didn't just follow the trail in modest numbers, they exploded out in a crowded column. They streamed back and forth and all along the trail I had laid. They were like people reacting to a very loud fire alarm in a crowded building, running out and about and shouting." As if making first contact with an alien life-form, Wilson had spoken to his ants, and they had heard him and responded.

Having evidently found the source of the pheromone that guided ants to their forage, Wilson next wanted to determine the substance's chemistry. "If successful, it would be a real breakthrough," he writes, "an ant word identified in the ant Rosetta Stone." A recent fortuitous development in chemical analysis, the work of chemists in the petroleum and tobacco industries, made that identification possible. It yoked two instruments that together could sort out small samples of mixed organic substances—substances such as the ants' trail pheromone. One of the instruments was the gas chromatograph. It vaporized a compound and ran the resulting gas mixture through a column that separated the components. The other instrument was the mass spectrograph, which electrified the gaseous components the chromatograph produced, spread them out according to their mass, and displayed the resulting mass spectrum as a graph. That picture of distinct chemical peaks could then be compared with the spectra of known compounds to determine what the components were.

But a small sample for chemical and physical analysis was still much more than the bare speck of gland Wilson had extracted from a single fire ant. Each ant could supply, at best, a few micrograms—millionths of a gram. He needed milligrams—thousandths of a gram. To assay the fire-ant trail pheromone, then, Wilson would have to collect tens of thousands of fire ants.

He was no longer working alone. Anticipating his need for colleagues skilled in chemical analysis, he had recruited three chemists who had learned the new assay techniques to join him in the summer's work: one from the Baylor College of Medicine in Houston, Texas; one from the University of Chicago; one a graduate research student at the Harvard Medical School.

Wilson knew where to find fire ants, by the hundreds of thousands, and he knew how to harvest them. In the late summer of 1958, he and his three chemist colleagues traveled south from Massachusetts to northeastern Florida, to a rural highway between Jacksonville on the coast and the Okefenokee National Wildlife Refuge inland. "Mound nests of fire ants abounded there," Wilson writes, "next to ponds fed by quietly moving freshwater streams." A typical mound might house as many as two hundred thousand ants, and there were mounds along the roadside, spaced apart about the length of a football field, as far as the eye could see.

The red imported fire ant, a native of the South American floodplain, is adapted to protect its colonies from inundation, Wilson explains:

> When the ants sense the approach of a flood from around and below them, they move to the surface of the nest, carrying with them all the young of the colony—the eggs, the grublike larvae, and the pupae—while nudging the mother queen upward as well. When the water reaches the nest chambers, the workers form a raft of their bodies. The whole colonial mass then floats safely downstream. When the ants contact dry land, they dissolve their living ark and dig a new nest.

By a rural Florida roadside on a summer day, the scientists broke out spades and buckets and began spading scoops of fire-ant mound into the buckets, which they dumped into a roadside pond. As the sand and dirt sank beneath the pondwater, the scrambling ants floated to the surface and assembled their living rafts. These the men scooped up with tea strainers and transferred into bottles they had partly filled with a low-boiling-point solvent, dichloromethane, ready to be distilled.

"Do what we can," Wilson's New England predecessor Ralph Waldo Emerson wrote in one of his essays, "summer will have its flies; if we walk in the woods we must feed mosquitoes." If they collected fire ants, they must suffer the stings. "We came back [to Harvard] with the requisite one hundred thousand worker ants (roughly estimated, not counted!)," Wilson reminisces, "and my hands covered with itching welts from the stings of many angry ants." His hands, and the hands of his colleagues. They never again volunteered to collect fire ants with Ed Wilson.

Now the chemists went to work teasing out the trail pheromone. They did that by steam-distilling batches of their solvent-dissolved ants, extracting the oily glandular secretions from the distillate with petroleum ether, another common laboratory solvent, and then separating the Dufour's gland component with gas chromatography. Wilson tried laying trails with this extract to see if the ants responded as they had in his earlier experiments that used the crushed gland itself. They did, he reported in the journal *Science* in March 1959; the extract produced "trail-following responses of nearly comparable magnitude to those produced by [crushed Dufour's] gland preparations."

Wilson was thus the first investigator in the world to identify the basic system whereby ants communicate, an original and significant achievement. Along with that work in 1958 and 1959, he published more than twenty scientific and popular scientific papers in journals as diverse as *Science* (the premiere U.S. scientific journal), *Psyche* (a leading entomological journal), *Pacific Insects, Natural History,* and *Scientific American.* It isn't surprising that he was awarded lifetime tenure and an associate professorship at Harvard in 1958, at the young age of twenty-nine. He had earned the distinction, however much it made Jim Watson swear.

Yet, despite his and his chemist colleagues' painful work of fire-ant collection and pheromone extraction—not to mention the tedious repetitiousness of running the trail-following experiments six and more times each, measuring and counting, keeping track of ants no bigger than bread crumbs—the experimental trail trailed off. The Dufour's gland substance proved to be "a relatively simple molecule," Wilson writes, "—a terpenoid—and its complete molecular structure seemed within reach." (Plants and insects produce a large variety of biologically active volatile oils known as terpenoids, many with intense odors, among them citronella, rose oil, and turpentine.) But the more the chemists purified the secretion, the less eagerly the ants responded. Something was awry.

"Was the pheromone an unstable compound?" Wilson asked himself. "Was the effect the result of not one but a mixture of substances being required?" He decided a mixture was a good possibility. His colleagues agreed, and, worse, saw no way to tease out the components of such a compound given the current state of their chemical-analysis technology. Wilson's blunt conclusion, which speaks to what had been, after so much work, real pain: "We quit."

Before they abandoned their research, they decided to sum it up in a report, a rare scientific paper describing a negative result, to establish a baseline for further investigation as chemical-analysis technology improved. The paper was sufficiently novel to win acceptance by the British journal *Nature*. Even there, although the trio acknowledged experiencing "some frustration" in the negative outcome of their attempts to identify the active trail substance, they saw a potential discovery in the fading of their pheromone. The fact that the purer extract failed to activate the ants, they wrote, "indicates the possibility that chemical instability may play a part in obliterating the trail substance after it has served its function. Wilson has previously suggested that this is accomplished by using a very volatile material for a trail substance." A trail marker that continued to trigger ants to follow a trail after a food source had been collected would misdirect workers to a dead end.

And, in fact, in further studies, Wilson found that signal pheromones fade quickly unless other workers reinforce them—in the case of alarm signals that harvester ants spray into the air, within about thirty-five seconds. "The advantage to the ants of an alarm signal that is both local and short-lived," he wrote in *Scientific American* in 1963, "becomes obvious when a [harvester ant] colony is observed under natural conditions. The ant nest is subject to almost innumerable minor disturbances. If the [sprayed] alarm spheres generated by individual ant workers were much wider and more durable, the colony would be kept in ceaseless and futile turmoil. As it is, local disturbances such as intrusions by foreign insects are dealt with quickly and efficiently by small groups of workers, and the excitement soon dies away."

Not until 1981 did a research team—headed by Robert K. Vander Meer, the director of the U.S. Department of Agriculture's fire-ant laboratory in Gainesville, Florida—tease out the complete mix of pheromones that combined in the ant Dufour's gland. "The trail substance," Wilson reports of that work, ". . . is not a single pheromone but a medley of pheromones, all released from the sting onto the ground. One attracts nestmates of the trail layer, another excites them into activity, and still another guides them through the active space [of vapor] created by the evaporating chemical streaks. All of the components need to be present to evoke the full response." Wilson calls this combination "engineering by natural selection." By not realizing its complexity, he concludes, "and thereby taking aim only at one of the components, we had failed to identify any of them." Because evolution operates by modifying existing traits rather than inventing new ones, few phenomena in the natural world are simple. Yet Wilson's contribution to the unraveling of the ant communication system was fundamental. "I was able," he recalled much later, "to at least be the first to piece together how pheromones were being used as a means of communicating information between ants and within a colony."

A more definitive study Wilson conducted in that productive 1958 year offers a rare example of scientific comedy. He had noticed a curious behavior that he decided to explore with two of his chemist colleagues: ants, which are fastidious in keeping their nests clean, left dead nestmates standing or lying where they had died for a day or two before picking them up and carrying them outside (under laboratory conditions, to a refuse pile they established well away from the nest). They even groomed a dead nestmate as if it were still alive, however crumpled and immobile it might be. Since ant behavior is mediated by odor, Wilson wondered why the odor of decay—"a pheromone of sorts," he notes—didn't trigger more prompt disposal of the corpse. "If live ants used pheromones to release other instinctive social behavior in the service of the colony," he speculated, "why not in death also?"

As he thought over the question, he realized that the ant's rigid chitin exoskeleton probably confined decay odors until decomposition had generated enough gas to force the odor out. (Chitin forms the shells of lobsters and crabs as well as of insects; tough and waterproof, it's also the primary component of fish scales.) Here was a simple, conspicuous, and stereotyped pattern of ant social behavior worth exploring.

Wilson did that with a laboratory colony of dark rust-red Florida harvester ants, *Pogonomyrmex badius,* which had established a refuse pile about a yard away from the nest entrance, at the far side of the colony's closed foraging area. (An external refuse pile was an adaptation to confinement; in the wild, Florida harvester ants simply dump their dead immediately outside their entrance crater, where scavenger species soon haul them away to feed on. Or, even more accommodating, sick and dying harvester workers leave the nest voluntarily and wander around ignored until they expire. As Wilson put it: "The removal of corpses is thus accomplished in large part by the dying ants themselves, as their last act as living workers.")

For the first of a series of experiments, Wilson's chemists made an acetone extract of worker corpses, which Wilson daubed onto five ant-sized squares of filter paper. Five more little squares, untreated, served as controls. All ten were arranged inside the foraging area near the nest entrance. In three separate trials, the first workers to come across the treated squares picked them up and hauled them over to the refuse pile. Except in one case, the workers left the untreated squares alone.

Wilson varied the same experiment with other materials. Harvester ants would ordinarily either ignore seeds or pick them up and take them into the nest to be stored; if the seeds were drizzled with corpse extract, they went off in the mandibles of diligent workers to the refuse pile. A pure extract of dead cockroach that Wilson tried made the laboratory "smell faintly of a mixture of charnel house and sewer," because two of the extract components, the terpenoids indole and skatole, are "elements of mammalian feces." Most of the non-ant substances he tested caused only excitement and aggressive circling, but the essence of dead cockroach with its fecal terpenoids triggered what Wilson had started calling the "necrophoric" response—workers carrying the drizzled samples, in this case bits of balsam, to the refuse pile.

Wilson was delighted with the success of his experiments. "There is no procedure more pleasing to a biologist," he writes, "than an experiment that works." He repeated his necrophoric baptisms for everyone who came into his laboratory until the repetition bored him.

The ultimate test, it occurred to him then, would be to see what his ant colony's tidy workers would do with a live ant drizzled with dead-ant juice. He thought of the experiment as turning ants into zombies, although these would be zombies in reverse—not the dead brought to life, but the living shrouded with the odor of death.

From an ant's perspective, Wilson's experiment might have seemed like a visitation from a god: a giant leaning over its nest, a vast transparent glass cylinder sweeping down, a drop of a smelly liquid washing over its body. "The result was gratifying," Wilson incants. "Worker ants that met their daubed nestmates picked them up, carried them alive and kicking to the cemetery, dropped them there, and left. The behavior of the undertaker was relatively calm, even casual. The dead belong with the dead."

Robert Frost anticipated Wilson's observations in the later lines of his tongue-in-cheek 1936 poem, "Departmental":

Ants are a curious race;
One crossing with hurried tread
The body of one of their dead
Isn't given a moment's arrest—
Seems not even impressed.
But he no doubt reports to any
With whom he crosses antennae,
And they no doubt report

To the higher-up at court.
Then word goes forth in Formic:
"Death's come to Jerry McCormic,
Our selfless forager Jerry.
Will the special Janizary
Whose office it is to bury
The dead of the commissary
Go bring him home to his people. . . .
And presently on the scene
Appears a solemn mortician;
And taking formal position,
With feelers calmly atwiddle,
Seizes the dead by the middle,
And heaving him high in air,
Carries him out of there.
No one stands round to stare.
It is nobody else's affair.
It couldn't be called ungentle,
But how thoroughly departmental.

Frost found ant mortuary behavior amusing enough when it involved dead ants; Wilson, toying with live ants as Jehovah and Satan toyed with Job (but more kindly), pushed it on to comedy. "After being dumped on the refuse pile," he recalls, "the 'living dead' scramble to their feet and promptly return to the nest, only to be carried out again. The hapless creatures are thrown back onto the refuse pile time and again until most of the scent of death has been worn off their bodies by the ritual." Since ants badly injured in the field are commonly carried home and eaten, merely conflating the smirched workers with the dead until they cleaned themselves seems benign.

Frost's allusion to formic acid refers to the earliest of ants' many volatile compounds to be identified. An English Restoration-era parson-naturalist named John Wray reported his observations on ant secretions in a letter to the *Philosophical Transactions of the Royal Society of London* in 1670. Wray called ants by their old name, "pismires," a name that reflects the common observation that an anthill smells like urine—a smell some noses confuse with formic acid. Wray reported: "A weak Spirit of Pismires will turn [blue] Borage flowers red in an instant: Vinegar, a little heated, will do the like. Pismires distilled by themselves, or with water, yield a Spirit like Spirit of Vinegar. . . . When you put the Animals [i.e., the ants] into water [to distill them], you must stir them to make them angry, and then they will spirt out their acid juyce. No animal that ever we distilled . . . except this, yields an *Acid* Spirit . . . and yet we have distilled many, both Flesh, Fish, and Insects."

After Wilson's 1958 discovery of the source of a trail pheromone component in the Dufour's gland, a number of researchers armed with increasingly sophisticated instruments sorted out most of the many pheromones that ants secrete. By 1983, Robert Vander Meer could summarize reports of no fewer than nine different chemical compounds distilled from Dufour's gland extract that contribute to trail marking alone, and in total, from just the Dufour's gland, more than fifty. By then, other pheromones produced elsewhere in the ant body had been identified as well: queen pheromones, brood pheromones, nestmate recognition pheromones, worker attractant pheromones, and more. Pheromones were ant language, a chemical dictionary of commands.

With them, Wilson would write, these highly social insects could even "put together pheromones with other odors to create 'proto-sentences.' A foraging worker, having encountered fire ants, rushes into its home nest with the equivalent of shouting *Emergency, danger* by spraying alarm pheromones, then *Enemies* by presenting the odor of fire ants it has acquired on its integument during a recent tussle, and then *This way, follow me,* as it turns and runs back along the odor trail it has just laid."

Wilson's fundamental experiments during this period of his career demonstrated that organismic biology could be as productive of discoveries as molecular biology and need not desiccate to a dry-bones molecular reductionism. At the same time, he had not forgotten his earlier work in the South Pacific and the questions it raised about how species colonize islands. Given the extremely complex and overlapping relationships among multiple species on continental landmasses, might island colonization serve as a simpler model through which to study how evolution proceeds when species encounter each other and adapt? And since islands were numerous and variously isolated, and came in all sorts of sizes and shapes, couldn't they serve for natural experiments as well, making it possible to test a theory under a variety of conditions? And wasn't an island something more than a landmass surrounded by water? Mountains were insular; so were stands of forest surrounded by prairie; so were streams, caves, and tide pools; so, most urgently, were the formerly continuous natural habitats throughout the world increasingly broken up by human settlement. Learning how species were faring under such conditions would be important to preserving the natural world while the human world continued, seemingly inexorably, to encroach upon it.

Just as he was beginning to think through this program of research, Wilson encountered a colleague whose original contributions to the question of how species make space for themselves in a crowded environment were stirring both excitement and controversy. The work that Ed Wilson and Robert MacArthur would accomplish together in the early 1960s in island biogeography would set the burgeoning field on solid ground. It would lead as well to Wilson's writing, with MacArthur, his first book.

6

Keys

ED WILSON FIRST ENCOUNTERED Robert MacArthur in 1960, looking to recruit him for the group Wilson had pulled together to fight the molecular wars and advance the cause of evolutionary biology. "By this time I had become radicalized in my views about the future of biology," Wilson recalls. ". . . I wanted a revolution in the ranks of the young evolutionary biologists. I felt driven to go beyond the old guard of Modern Synthesizers and help to start something new." He didn't yet know how to make that happen, but the first step was clearly to recruit fresh talent.

Earlier that year, a publisher had asked Wilson to evaluate a manuscript by a young marine biologist, Lawrence Slobodkin, then teaching at the University of Michigan. "It dawned on me that ecology had never before been incorporated into evolutionary theory," Wilson recalls his response. The manuscript, which became Slobodkin's 1961 book *Growth and Regulation of Animal Populations,* showed a way to do so. Wilson returned an enthusiastic report. Then he approached Slobodkin to propose that they write a textbook together: to advance a field, write a new textbook.

Wilson and Slobodkin pursued their planning at the 1960 annual meeting of the American Association for the Advancement of Science (AAAS), held between Christmas and New Year's in New York City. Both were participating in a Tuesday two-session symposium on "Modern Aspects of Population Biology." Population biology is the study of the biology of whole populations, rather than, say, of individuals or energy flows. Since it deals with patterns of population change, it's necessarily a field heavy with mathematics, never Wilson's strong suit. He and Slobodkin needed a mathematician to help with their textbook. Robert MacArthur, who had earned both undergraduate and master's degrees in mathematics, might be that man. Both Wilson and Slobodkin were presenting papers in the Tuesday symposium, as was MacArthur, and his would be the last session of the day. Slobodkin knew the mathematician-biologist. They'd done their doctoral work at Yale at the same time, under the same faculty adviser. "He's a real theoretician," Wilson recalls Slobodkin praising MacArthur—"very bright." They attended MacArthur's AAAS session together and waited for him afterward. Wilson had never met him before.

Wilson and MacArthur found immediate connection. Among other affinities, they were similarly entrepreneurial in their approach to doing science. "Robert (he resisted being called Bob) and I were relatively young," Wilson writes. "He was twenty-nine and I was thirty. We were both very ambitious, each searching self-consciously for the opportunity to make a major advance in science." Slobodkin's "very bright" understates MacArthur's reputation. Wilson held his own with him, but MacArthur awed and intimidated many of his contemporaries. "He was widely thought to be the new avatar of theoretical ecology," Wilson says, "having already made several seminal advances. He was an avid naturalist and expert on birds, and in addition (very important in our case) an able mathematician." That was the context of his contemporaries' awe. But, "thin, sharp in face and disposition," Wilson adds, with "an intense and withdrawn manner that warned off fools, [MacArthur] was not the kind who placed hand on shoulder and slapped backs, nor did he often laugh out loud." They would work together for most of the next decade, work that would culminate in writing a book together, but they never became close friends. "We never finished taking the measure of each other," Wilson concludes.

MacArthur was born in Canada to American parents. His father, John Wood MacArthur, was a professor of genetics at the University of Toronto at the time, 1930; his mother, Olive, was a bacteriologist and botanist. The MacArthurs were outdoor people who owned a summer home east of the Green Mountain National Forest in southern Vermont. Young Robert explored and collected across summers there much as Wilson had in rural Alabama, but with the stability of a solid family behind him.

In 1947, when he was seventeen, Robert learned of the impending opening of a small liberal-arts college in Marlboro, Vermont, near his family's summer home. Marlboro College would be an egalitarian institution, with no faculty rankings and no grades, with a town-hall government, where students could work independently and chart their own course. Its founder, Walter Hendricks, a professor of humanities at the Illinois Institute of Technology who owned a farm nearby, was a lifelong friend of the poet Robert Frost, who had encouraged and supported the new venture. A news story in *Time* magazine announced the school's opening, and MacArthur's physicist older brother, John Junior, remembers that his brother enrolled in the small opening class, in response to the *Time* story. Since the MacArthurs summered nearby, Robert probably knew of the school from the extensive remodeling and construction ongoing that 1947 summer on the farm Hendricks had purchased next to his own for a campus. Once in residence, his brother recalled, Robert wrote his family "and told us what it was like. We decided we trusted his judgment, so we quit our jobs and came [the following year]." Quitting their jobs is a remarkably casual way to describe John Senior's leaving behind a tenured position for an uncertain future at a new college opened on a shoestring. John Junior moved to Marlboro to teach physics, and when John Senior died, in 1950, Olive MacArthur stepped in to teach biology. Marlboro College became something of a family enterprise.

Robert MacArthur thrived at Marlboro, majoring in mathematics and continuing his birding in the woods of southern Vermont. After graduating from Marlboro in 1951, he earned a master's degree in mathematics at Brown University in Providence, Rhode Island. Wilson would later judge him to be an exceptional but not a first-rank mathematician, as almost no scientists are. They don't need to be, Wilson points out, and if they were, they would go into mathematics. Which may explain why MacArthur sought admission to a Ph.D. program not in math but in ornithology—at the University of Illinois, under the ornithologist Charles Kendeigh, who had published studies of bird communities in Northern forests. "In a committee decision that likely shaped the future of ecology," writes one of MacArthur's eulogists, "Illinois rejected his application, ostensibly because he was trained as a mathematician, not a biologist." He tried again, this time in zoology, this time at Yale, and was admitted to graduate study under G. (for George) Evelyn Hutchinson, a polymath British limnologist—limnology is the scientific study of lakes—who was in the process of founding the science of ecology. (I knew Hutchinson. He took afternoon tea several times a week at the Elizabethan Club, a Yale institution; a fellow member, I sometimes joined him. I wish I had known him better, well enough to ask him about his extraordinary life and work. MacArthur was also a graduate student at Yale when I was an undergraduate, in the late 1950s, though we never met.)

Between degrees, the military draft intervened. MacArthur spent two years in the army at Fort Huachuca in Arizona, calculating artillery trajectories. He put his spare time to good use, birding, reading, and doing research, including planning his doctoral thesis. "He would later tell colleagues," the eulogist reports, "that, upon returning to New England, he had the basic story of [his thesis research] after two weeks in the field." He may have, but he needed the summers of 1956 and 1957, and a between-term winter trip to Costa Rica, to support and document the work.

MacArthur's doctoral dissertation, "Population Ecology of Some Warblers of Northeastern Coniferous Forests," became famous for its rigorous demonstration of niche theory, a concept his mentor, Evelyn Hutchinson, was just then in the process of redefining formally and mathematically. (Loosely, a niche is a species' place in the environment.) It asked and answered the question: when similar species of birds occupy the same areas of forest, what prevents them from competing for resources until all but one of them are extinct?

The young ornithologist answered the question by demonstrating that the several species of warbler he studied, all of which hunted and fed on insects, essentially divided up the living zones of the mature evergreen forest trees they shared. So Cape May warblers fed mostly in the top third of the sixty-foot Maine spruce trees, and in the outer zone of that foliage. The black-throated green warbler, in turn, fed mostly in the middle third of the spruces, in both the outer zone and the next zone inward, toward the trunk. The myrtle warbler fed in the more open bottom third of the trees, an area that included both the middle and inner zones and the open space around the trunk below the lower foliage, down to the ground. (For clarity, I've left out two other, less numerous species that MacArthur also investigated, with similarly zoned behaviors.)

It should follow that each species preferred different prey, prey that in turn preferred the different zones of the trees where the warblers hunted. MacArthur showed that the warblers' preferences were in fact for different species of insects common to their preferred zones. The warblers might differ as well in their feeding behavior, and they did. Among a number of different feeding behaviors MacArthur catalogues: different species moved in different directions as they hunted—vertical, tangential, or radial—made more or fewer flights to other trees, spent more or less time "hawking," meaning flying off after insects, and varied in their energy levels from active to sluggish. In these and other ways, the several different species of warblers MacArthur studied staked out spaces for themselves—niches—that minimized their competition and allowed them to coexist more or less stably in the same trees.

What was different about MacArthur's warbler study was its strategy: rather than simply describing the natural history of the warblers and their interactions, it formulated theories and then tested them against the real world of real birds in a real forest. It counted and compared and quantified its findings and drew further generalizations from them that could be tested in their turn. As a fellow ecologist wrote at a later time, "The arguments MacArthur's work engendered, and the way they changed ecology, illustrate science as a process, not an encyclopedia."

With his Ph.D. awarded, MacArthur went off to Oxford for a year to deepen his knowledge of ornithology under the evolutionary biologist David Lack, whose 1947 book, *Darwin's Finches,* neglected when it was published, was just being resuscitated for its insights into species competition as a driver of evolution. When MacArthur returned from England, he took up teaching as an assistant professor at the University of Pennsylvania. It was then that Wilson and Slobodkin recruited him for their textbook team.

That project eventually foundered on differences in approach and temperament among the three men. But MacArthur enthusiastically joined the cadre of evolutionary biologists Wilson and his colleagues were assembling to push their field into more experimental, rigorous, mathematically informed science. Their association culminated in a gathering in the summer of 1964 at the lakeside MacArthur family home outside Marlboro. Besides Wilson and MacArthur, the group of five included another Hutchinson protégé, the tropical ecologist Egbert Leigh; Richard Levins, a theoretical population biologist who would later join the faculty of the Harvard School of Public Health; and Richard Lewontin, a pioneer in theoretical and experimental genetics who would assume a Harvard professorship in zoology a decade later and eventually chair Wilson's department. Slobodkin was unable to attend, but Wilson kept him informed.

In *Naturalist,* Wilson's 1994 memoir, he recalls the gathering nostalgically:

> For two days between walks in the quiet northern woodland, we expanded upon our common ambition to pull evolutionary biology onto a more solid base of theoretical population biology. Each in turn described his particular ongoing research. Then we talked together about the ways in which that subject might be extended

> toward the central theory [i.e., Darwinian evolution] and aligned with it.

They decided to pool their work, to write a series of "audacious and speculative" essays under a common pseudonym to advance their cause. If that decision sounds like a midsummer night's dream, it was. Nothing came of it. "So the program faded," Wilson writes, "and for the most part the conspirators went their separate ways. We never met as a group again."

It isn't surprising that so driven and ambitious a group of young scientist-academics, however moonstruck, would soon turn back to their own careers. For Wilson at least, the gathering was a watershed. "I cannot speak for the others," he recalls of that 1964 summer, "but I believe we all carried away a new confidence in the future of evolutionary biology, and in ourselves." Some of Wilson's detractors in graduate school had disdained the earnest seriousness he'd brought with him from Alabama and nicknamed him "the growing boy." The growing boy was mature and confident now, and moving to the center of his profession.

Another event in the summer of 1964 further eased the conflict between the belligerents of the molecular wars. Theodosius Dobzhansky, born in the Ukraine in 1900, the evolutionary biologist who formulated the Modern Synthesis in an influential 1937 book, spoke that August at the annual summer meeting of the American Society of Zoologists in Boulder, Colorado. As the society's outgoing president, he had decided to weigh in on the question of molecular versus evolutionary biology (he called it "organismic biology," preferring his adjective to Wilson's "evolutionary"). He promised that what he had to say "may seem provocative to some of you; I do not expect it to seem boring."

Nor was it. Dobzhansky was blunt:

> Nothing succeeds like success. In molecular biology, one spectacular discovery has followed closely on the heels of another. Molecular biology has become a glamour field. It has attracted

> many able young students as well as older investigators. Glamour and brilliance generate enthusiasm and optimism; they may also dazzle and blindfold. The notion has gained some currency that the only worthwhile biology is molecular biology. All else is "bird watching" or "butterfly collecting."

The fundamental mistake of that attitude, Dobzhansky went on, was to ignore the levels of integration in the objects biologists study. "The opinion forcibly expressed by some molecular biologists is that, to be 'modern,' or even 'scientific,' organismic biology must be reduced to molecular biology. All that this means in most cases is that many molecular biologists are so excited about what they are doing that they are unable to see why their organismic colleagues can find excitement in something else."

Some of those enamored with the new discoveries, Dobzhansky cautioned his audience, urged the reduction of organismic biology to the molecular level because chemistry and physics are supposed to be more exact, more "advanced," and hence superior. But to the contrary:

> A gene . . . is a double-stranded DNA molecule, or perhaps a part of such a molecule. A chromosome is, however, not a heap of genes, but a configuration of genes arranged in a certain way which proved to be adaptively advantageous in evolution. A cell is not a conglomeration of chromosomes but a supremely orderly contrivance consisting not of chromosomes alone but also of many other organelles. An organ and an individual body are, in turn, not simply piles of cells but beautifully designed and often highly complex machines, in which the cellular components are not only diversified but often have lost their separate identities.

And so on, up through ecosystems, the largest biological system—the biological organization of entire populations. Both approaches to biological knowledge, Dobzhansky concluded, the molecular and the organismic, and probably others as well, were required to understand the deep complexity of the natural world, and of humankind in particular.

And just as the evolutionary biologists' moonstruck summer plans for a campaign would lose force in the months and years to come as real research intervened, so also, Wilson writes, "the molecular wars subsided to their ambiguous conclusion." The departmental contretemps at Harvard found its resolution in division, the old biology department with its fiefdoms splitting into departments of molecular biology on one side and of organismic and evolutionary biology on the other.

A young assistant who would become one of Wilson's secret weapons joined the biology team late in 1965. Kathleen Horton, then twenty-three, applied for work on a tip from a friend: Wilson's postdoc Robert Taylor, a New Zealander, was organizing to move to Australia to become curator of ants at the National Insect Collection in Canberra and needed someone to pin his large ant collection before the move. Though she was born in Orange County, California, Horton's background was international. Her parents were Baptist missionaries who had walked out of Burma (Myanmar today) in 1942 ahead of the Japanese invasion. She had grown up in Pakistan and attended her final two years of college in Lebanon. Taylor had warned her not to let Wilson know she could type, fearing he would take her away from ant pinning to type his papers. Somehow he found out. She finished Taylor's collection while beginning work with Wilson; she was still working with him in 2021. She has kept him organized and interfaced with his students and visitors. Wilson writes all his books and papers in longhand on yellow legal tablets. Horton types them, nearly four hundred of them across the years. "I went through four typewriters and six computers," she told me of her work.

Wilson's encounters with Larry Slobodkin and Evelyn Hutchinson led him to explore the synthesis now known as evolutionary ecology: "Through them I came to appreciate how environmental science might be better meshed with biogeography and the study of evolution, and I gained more confidence in the intellectual independence of evolutionary biology. I was encouraged to draw closer to the central problem of the balance of species, which was to be my main preoccupation during the 1960s." Biogeography is the branch of biology that deals with the geographical distribution of plants and animals. Since their first meeting at the AAAS in New York in 1960, Wilson and MacArthur had been exploring together how to develop an *experimental* biogeography, focused on islands. "We both saw ecology and evolutionary biology as potentially one continuous discipline," Wilson recalls, "filled with opportunity for innovation in theory and field research."

In an experiment, the experimenter tries to arrange conditions so that only one factor changes at a time. "Theoretically," MacArthur writes, "if more than one factor changes, the analysis can still be performed, but in practice, if more changes of known nature occur, more of an unknown nature usually also occur." So, for example, if they introduced a new species to a small island, they could see how that introduction affected all the other species occupying that island. Introducing more than one new species would make it difficult to sort out which new introduction caused which changes. There might be complex interactions between the newly introduced species as well.

"I spoke to MacArthur about islands I had visited around the world," Wilson recalls, "and their use in studying the links between the formation and geography of species. I could see that he was not thrilled by the complexity of the subject." MacArthur showed much more interest in Wilson's graphical plots than in his narratives; the graphs generalized the subject into numbers MacArthur could see how to manipulate. Wilson had been plotting species-area curves, he writes:

> These displayed in a simple form the geographic areas . . . of islands in different archipelagoes of the world, principally the West Indies and western Pacific, and the number of bird, plant, reptile, amphibian, or ant species found on each island. We could see plainly that with an increase of area from one island to the next, the number of species increased roughly to the fourth root

> [i.e., the square root of the square root]. This means, for example, that if one island in an archipelago is ten times the size of another in the same archipelago, it would contain approximately twice the number of species. We also observed that islands more distant from the mainland had fewer species than those close by.

What they found interesting about these differing distributions was the way the immigration of new species seemingly came to match the extinction of settled species, with the two lines on their graphs crossing at points of larger or smaller total population, depending on an island's size and its degree of isolation. "On small islands," Wilson writes, "the crowding of the species is more severe, and the extinction rate curve is steeper. On distant islands, immigration is less, and the immigration curve less steep. In both cases the result is a smaller number of species at equilibrium"—that is, the number of species holds more or less constant, although the specific mix may change. Wilson called this phenomenon "saturation."

The two young biologists pulled their ideas and data together into a landmark paper published in the journal *Ecology* in 1963. Though the paper was little noticed at the time, Wilson says, it came to be considered a landmark in the new field of ecology, with a graph that would be reproduced routinely in textbooks for decades after:

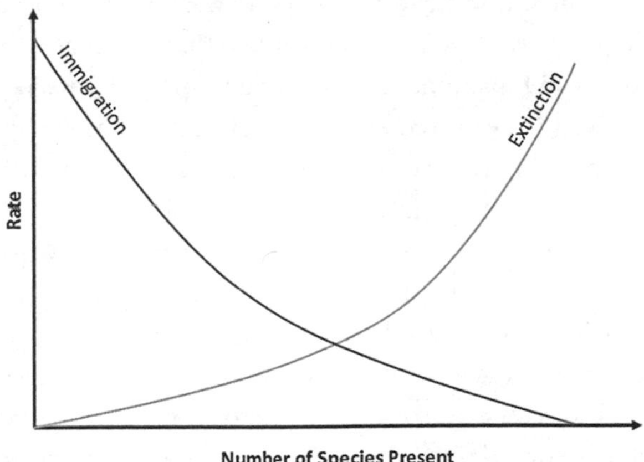

FIG. 4. Equilibrium model of fauna of a single island. See explanation in the text.

Here, in vivid form, was a representation of the equilibrium model that Wilson and MacArthur had discerned in the interlocking patterns of immigration and local extinction of species colonizing a new island. A more elaborate graph in the same paper visualized how the point of equilibrium varied with island size and with the distance of an island from the mainland.

In effect, the graphs made predictions about island colonization that could be tested by visiting such islands and actually identifying and counting species and then following the populations to see how colonization and local extinctions proceeded. If they matched the predictions of the graphs, Wilson and MacArthur would have discovered a basic pattern of ecological increase and decrease that might have application in the larger world—for example, in understanding how to harvest forests sustainably without depopulating the habitat.

There are many forms of "islands" in the world, a later observer notes; Wilson and MacArthur's equilibrium theory eventually "attracted great interest because its predictions were not confined solely to 'proper' oceanic islands; the theory was applicable in any situation where patches of habitat favourable to the 'island' species were separated from one another by areas of unfavorable habitat." Unfavorable habitat could be an ocean, a river, a gorge, or even, reversing field and ground, land separating lakes.

Jared Diamond, in his early scientific work prior to becoming a best-selling author, noted in a 1975 paper the challenge of what he called "the island dilemma" for the design of nature reserves. When natural habitat is destroyed by logging or cultivation, he wrote, the size of a remaining reserve and its location in relation to other such reserves is important: "If most of the area of habitat is destroyed, and a fraction of the area is saved as a reserve, the reserve will initially contain more species than it can hold at equilibrium. The excess will gradually go extinct. The smaller the reserve, the higher will be the extinction rates. . . . Different species require different minimum areas to have a reasonable chance of survival." Diamond also applied the lessons of the Wilson-MacArthur model to propose better arrangements and geometries of reserves to protect wildlife.

With a surprising lack of response from the biology community to their 1963 paper, Wilson and MacArthur decided in 1964 to expand the paper into a book about these seemingly universal patterns, trading draft chapters back and forth for the next several years. "We applied this simple [saturation] model to every scrap of data we could find," Wilson recalls, "on related subjects in ecology, population genetics, and even wildlife management, and fitted it together as best we could."

Another development in Wilson's life marked the early 1960s. He and Irene decided they were sufficiently established to raise a family. Irene's health had declined since the early days of their courtship, so much so that her frailty precluded bearing children. Instead, the Wilsons adopted a baby daughter, Catherine, in 1963. For the next decade, Cathy and Irene would join Ed on his summer excursions, to their mutual delight.

In the autumn of 1964, Wilson became graduate adviser to a talented young biologist, a recent Harvard graduate, Pennsylvania-born, named Daniel Simberloff. Like MacArthur, Simberloff had trained as a mathematician, though he'd made the switch to biology sooner, in his undergraduate senior year. He crammed biology that year, but his training was still spotty when Wilson took him on. "I was concerned about getting into grad school," Simberloff told me. "I hadn't had that much biology." A friendly faculty member encouraged him to speak to Wilson. "I went up and talked to Ed, and he made a deal with me. The deal was that he would teach me all the biology that I'd missed and that I needed, and I would help him deal with math and teach him some math."

The first thing Wilson did when Simberloff became his grad student was to hand his new charge a recently published paper on kin selection, a stunning but heavily mathematical and largely ignored work by a young British evolutionary biologist named W. D. Hamilton. In that paper, "The Genetical Evolution of Social Behaviour," Hamilton demonstrated the evolutionary fitness of behaviors that classical Darwinian theory lacked an explanation for, such as a parent's willingness to sacrifice himself or herself for their children: "In certain circumstances an individual may leave more adult offspring by expending care and materials on its offspring already born than by reserving them for its own survival and further fecundity." Hamilton then explored what those circumstances might be, and proposed that this "kin selection," based on the percentage of genes the individuals shared, best explained them. Full siblings—brothers, sisters—share 50 percent of their genes. Half-siblings with only one common parent share 25 percent of their genes. First cousins share 12.5 percent of their genes. Depending on how closely related they were, more relatives would have to be involved in support to pass along the same percentage of genes. As Hamilton famously framed it: "To express the matter more vividly . . . we expect to find that no one is prepared to sacrifice his life for any single person but that everyone will sacrifice it when he can thereby save more than two brothers, or four half-brothers, or eight first cousins." Here was an expansion into social behavior that diluted the pure "selfishness" of classical Darwinism. Wilson's curiosity about Hamilton's work was an early signal of his growing interest in social biology.

Wilson also gave Simberloff a copy of an early draft of the book he and MacArthur were writing. "He asked me to comment on it," Simberloff says, "and especially to read the math chapters that Robert had written. I read it, and I made a bunch of comments. I said: You know, it all fits very well, but there's not really direct evidence for extinction. You're citing relationships which could be explained by your equilibrium theory, but they could be explained by other phenomena as well. His response was: Well, why don't you test it?"

Simberloff didn't say whether he was stunned by Wilson's proposal—a new graduate student, confident of his math abilities but unsure if his biology was up to muster, asked to test a theory his brilliant mentor and another brilliant biologist were writing a book about. Whatever his immediate reaction, Simberloff took up the challenge. "Ed didn't say anything about the Florida Keys at that time. I spent a while looking for a place to set up a test. I had spent time on the coast of Maine—just hiking, you know. I'd noticed these small islands there. So I went up there and found that they had beetles, lots of beetles, ground beetles, carabids. I thought maybe that could be a good site. But by October it became apparent that it was not going to be feasible to work year-round on an island in the Gulf of Maine."

By then, early 1965, Wilson had been thinking about how to clear an isolated area of all its wildlife, so that a test of the equilibrium model could start clean. Studying U.S. Geological Survey maps of the United States, he'd looked for islands along the coasts and even in lakes that might be suitable. Florida was on his mind, whether he mentioned it to Simberloff or not, because he and MacArthur had been discussing setting up a biological field station in the Florida Keys where Wilson could move his family to live and work during university vacations.

Wilson's first idea for testing the model had been to conduct a natural experiment: to find a location where nature had defaunated an island for him. In the 1963 paper he wrote with MacArthur, the two had discussed one such natural experiment, the August 1883 eruption and partial destruction of the island of Krakatau in the Sunda Strait—northwest of Australia, between Java and Sumatra—an eruption so massive that its ash and smog spread worldwide, dropping global temperatures by half a degree Celsius in 1884 and darkening skies for years afterward. The blood-red sunset in Edvard Munch's 1893 painting *The Scream* has been attributed to the vivid red sunsets Krakatau's pollutants produced. "Half of Krakatau disappeared entirely," Wilson and MacArthur write, "and the remainder, together with the neighboring islands of Verlaten and Lang, was buried beneath a layer of glowing hot pumice and ash from 30 to 60 meters thick. Almost certainly the entire flora and fauna were destroyed. The repopulation proceeded rapidly thereafter."

Obviously, the current experiment couldn't wait for the next volcanic eruption, but it occurred to Wilson that hurricanes probably swept some Florida islands clean. "My first idea," he told me, "was to go down to the Dry Tortugas, a chain of little sandy islands off Key West, to survey and map every plant and animal on them and then wait for a hurricane to wipe them clean. Then I would go back and study them. We actually got that started."

Wilson and Simberloff depopulated and studied Florida mangrove keys to see how they repopulated. Here, the smallest and largest mangrove keys in their study. Figure at far left in upper image is Simberloff, figure center in lower image is Wilson.

In 1966, Wilson took along a group of graduate students and systematically surveyed the Dry Tortugas—waterless, turtle-inhabited islands clustered about seventy miles west of Key West that culminate in Fort Jefferson, a massive fortress built across almost thirty years, from 1846 to 1875, to guard the shipping lanes entering the Gulf of Mexico. Two hurricanes blew through the Keys that 1966 season, Alma in June and Inez in October; Alma passed west of Key West, directly across the Tortugas. The double blow demonstrated the futility of relying on hurricanes to clean the islands under study, only to have them blown clean again within a few months. "I realized that wasn't going to do it," Wilson concludes, "so I had to figure out some other way of eliminating all these little arthropods." (The phylum Arthropoda includes invertebrates such as insects, spiders, crustaceans, and trilobites.)

He was already figuring it out that summer. After he and his students had surveyed the Tortugas (and weathered the early-June hurricane), Wilson had joined Simberloff in an area about fifteen miles up the chain of keys from Key West, northwest of Sugarloaf Key, where miniature keys of red mangrove, sometimes only one lone mangrove tree, rose out of the shallow waters at various distances offshore. Roughly circular, they were typically thirty to sixty feet across and fifteen to thirty feet high, and enough of them populated the area to allow for replicating an experiment at various distances from their source areas, the much larger keys along the chain down from the Florida mainland.

Simberloff recalls examining these small mangrove keys with Wilson that summer, breaking open twigs to find ant nests. "Some we could wade to," he told me, "some we had to take a boat to. We quickly determined that they had many of the ants Ed had seen in the fringing swamps." Not only ants but other insects as well. As they wrote in their subsequent paper:

> Breeding fauna of islands this size consists almost entirely of species of insects and spiders. . . . About 75 insect species (of an estimated 500 that inhabit mangrove swamps and an estimated total of 4,000 that inhabit all the Keys) commonly live on these small islands. There are also 25 species of spiders . . . and a few scorpions, pseudoscorpions, centipedes, millipedes, and arboreal isopods [i.e., small crustaceans such as wood lice and pill bugs]. At any given moment 20–40 species of insects and 2–10 species of spiders exist on each of the islands.

After obtaining permission from the National Park Service for their experiments, Wilson chose six small islands for their first series, each about forty feet in diameter, ranging in distance from the nearest source of potential colonists from 1,750 feet down to a mere seven feet. One of the six would remain untreated, to serve as a control. The others they would defaunate. Wilson had promised the resident National Park Service ranger not to kill the trees.

They first tried an insecticide spray, a mixture of parathion, diazinon, a sticking agent, and fresh water, spraying two of their experimental islands on 9 and 10 July 1966 until the mangroves were dripping wet. A day later, all the surface and bark denizens were dead, but they found a wasp, several small nests of ants, and two longhorn-beetle larvae still alive. A more thorough inspection nine days later turned up more ants and beetle larvae. Which meant a spray wasn't going to work. They would have to fumigate, somehow tenting their islands over to contain the fumigant, much as houses are tented to be fumigated against termite and other infestations.

Simberloff began calling exterminator companies around Miami. Wilson manned the phones as well, to explain what they were doing. "The first two or three we called," Simberloff remembers, "told us, 'Oh, we can't do anything like that. We fumigate houses.' But then one of them said, 'You ought to call Steve Tendrich at National Exterminators.'" They did, and Tendrich was interested. "He clearly viewed this as a challenge and different. We met with him and showed him what we needed. Then the experimenting began, both at his workshop in Miami and out on the islands."

How do you tent an island? Tendrich first tried using conventional steel scaffolding, laying down planks around the perimeter of the islands for the scaffolding to stand on to prevent it from sinking into the mud, adding a first tier of scaffolding to hold down the planks, then building up the open walls, one wall at a time, until the island was completely contained within a cube of scaffolding. Tendrich's men then used block and tackle to winch up the six-hundred-pound house-shaped fumigation tent and lower it into position over the scaffolding.

Wilson had found little information about fumigating live plants; fumigation is normally used to clear buildings, with living occupants, animal or vegetable, removed to safety. Most fumigants are soluble in water as well, meaning they would immediately clear the tent through the water and possibly concentrate in the water under the island—undesirable outcomes in either case. To identify a more effective fumigant, Tendrich ran field tests on red mangroves in a park in Coral Gables using a whole series of relatively water-insoluble chemi-

Exterminator Steve Tendrich first tried building a scaffolding cube around an island, then raising a tent with block and tackle and lowering it to cover the island for fumigation. The system worked but was cumbersome.

cals: methyl bromide, acrylonitrile, carbon tetrachloride, ethylene oxide, sulfuryl fluoride. Methyl bromide caused the least damage to the trees while reliably killing all the mangrove inhabitants, including roach eggs and butterfly pupae. It became the fumigant of choice.

Their troubles continued. They fumigated their first two islands during the day, when it was hot, and between the heat and the heat-stimulated chemical activity the mangroves were badly damaged. That meant they would have to fumigate at night, under lights. Methyl bromide is odorless; for their safety, they added chloropicrin—tear gas—to the mix as a tracer, to make sure they would know if they were risking exposure. The chloropicrin also proved to be effective at driving beetles out of deep boreholes in the mangroves. Finally, on 10 October 1966, in the middle of the night, they successfully defaunated the island called Experimental-7, E7, just offshore, at Bay Point in Manatee Bay, near Key Largo.

E7 was close enough to shore to minimize the problem of transporting the fumigation structural supports by small boat through the shallow water. The other experimental islands were located farther out, making the scaffolding cube system an awkward burden to

Steve Tendrich devised a more portable central-pole and guy-wire fumigation structure for the Wilson-Simberloff island experiments conducted in the Florida Keys.

move. At Christmastime that year, Tendrich discovered a more portable alternative. "He was driving around Miami," Simberloff recalls the story, "and he noticed a steeplejack setting up a tower on top of an office building to string Christmas lights from. He parked and took the elevator up to the top floor and walked up to the roof and introduced himself to the steeplejack. Told him what he had in mind and hired him on the spot."

What Tendrich had in mind was setting up a collapsible thirty-foot steel tower in the middle of one of Wilson's islands, driving it down through the mud into the mangrove roots and the underlying dead coral, stringing three guy wires out beyond the island anchored with mud screws, winching up the fumigation tent and then simply unfurling it down around the guy wires. He kept the tent edges submerged with sandbags and stakes. Wind was a problem—the tent couldn't be unfurled in winds greater than eight knots—but the wind actually helped when they needed to refurl the tent, pushing under the tent sides, lifting it up over the trees, and dropping it into the water.

Shark troubles occupied both Simberloff and Wilson. The two men remember the shark challenge differently. Wilson says the sharks were ground sharks, a small and harmless species that feeds on mollusks and shellfish. Simberloff says they were nurse sharks and blacktips, small but dangerous; blacktips account for about 16 percent of shark attacks annually in the Florida Keys. Tendrich was wary of sharks, but the steeplejack he'd hired was terrified of them. Both Wilson and Simberloff stood shark guard throughout the experiments, wading waist-deep in the shoals around the islands, armed with boat oars, knocking any approaching sharks on the head to warn them off. "During my entire doctoral dissertation work," Simberloff told me ruefully, "and two other projects I did after that using Tendrich's company, I was constantly dealing with sharks."

Through March and April 1967, the Wilson-Simberloff team fumigated their experimental islands, one after another. After a two-hour campaign of fumigation, they would open the tent seams for forty-five minutes to let the gas escape, remove the tent, and immediately spend several hours searching the fumigated mangroves for organisms. The methyl bromide did its work; they found "dead individuals . . . frequently in large numbers and in all life stages." Rarely were any organisms left alive, and those typically died within hours.

Once an island had been defaunated, it needed to be observed regularly for a year to follow its repopulation. That work fell primarily to Simberloff, who moved to Florida for the duration. He and Wilson were extremely cautious about contaminating the defaunated islands, allowing no boats to contact them and only seldom tying up to them; usually, they anchored at least forty feet offshore and waded in. They sprayed themselves with insect repellent before going ashore, and sprayed their equipment weekly with insecticide. They censused each of their six islands for two days every eighteen days, doing as little as possible to disturb the habitat as insects, spiders, and other small organisms began to colonize their miniature Krakataus.

At the conclusion of this immense scientific labor, in March 1969, Wilson and Simberloff published a two-part paper in a leading scientific journal, *Ecology,* reporting how they conducted their experiment and what they had discovered. They summarized their findings in an abstract of the second part of their paper:

> We report here the first evidence of faunistic equilibrium obtained through controlled, replicated experiments, together with an analysis of the immigration and extinction processes of animal species based on direct observations. . . . By 250 days after defaunation, the faunas of all the islands except the most distant one ("E1") had regained species numbers and composition similar to those of untreated islands even though population densities were still abnormally low. Although early colonists included both weak and strong fliers, the former, particularly the [tree-dwelling bark lice], were usually the first to produce large populations. Among these same early invaders were the [population groups] displaying both the highest extinction rates and the greatest variability in species composition on the different islands. Ants, the ecologi-

cal dominants of mangrove islands, were among the last to colonize, but they did so with the highest degree of predictability.

The most important finding of their research, they reported, was that its evidence supported the equilibrium theory Wilson and MacArthur had proposed first in 1963 and then, in more depth, in the book they wrote together, which had been published in the midst of the Florida Keys experiments in 1967, *The Theory of Island Biogeography.* Species colonizing a new habitat and species abandoning that habitat—going locally extinct—came to equilibrium at some point appropriate to the environmental conditions of the habitat. First there was little interaction, then a gradual decline in numbers as interaction (competition, predation) became important, and, finally, a lasting dynamic equilibrium, new colonists arriving, old colonists leaving or dying out, the mix of species changing, but the number holding steady as long as the environment did.

Simberloff had challenged Wilson (and MacArthur) to find experimental support for their theory. Wilson in turn had challenged Simberloff to help him mount an experiment testing the equilibrium theory. They had done so, together, and found confirmation. Their papers would become classics of the burgeoning new field of ecology, republished in 1991 in the collection *Foundations of Ecology: Classic Papers with Commentaries* along with the work of such fellow pioneers as Evelyn Hutchinson, Paul Ehrlich, and Robert MacArthur.

Besides its contribution to basic science, the work done by Wilson, MacArthur, and Simberloff has practical value in conservation biology as well. "Around the world," Wilson writes, "wild lands are being increasingly shattered by human action, the pieces steadily reduced in size and isolated from one another. Nature reserves are by definition islands. The theory serves as a useful tool in conceptualizing the impact of their size and isolation on the biodiversity they contain."

Dan Simberloff was awarded his Ph.D. in 1969. He accepted an offer of an assistant professorship at Florida State University, in Tallahassee, so that he could continue research in the Everglades and the Keys. Across a long career, he published more than three hundred scientific papers and some ninety book chapters, specializing in invasion biology, a field with echoes of his early work with Wilson on island colonization: "The patterns displayed by species introduced outside their geographic ranges," as he described it to me, "the impacts such species have on the communities they invade, and the means by which such invasions can be managed." In 1998, after his years at Florida State, Simberloff moved to a professorship in environmental science at the University of Tennessee, where he continues to investigate biological invasions, focused most recently on Patagonia.

When I spoke with him in the spring of 2020, he recalled his time working with Wilson with undiminished admiration for an unusual mentor. "It was really quite remarkable," Simberloff told me. "It was like taking five courses at once. The guy knew everything. It wasn't just insects and biogeography, but he knew about birds, he knew about fishes. We would talk about lots of interesting ecological and biogeographic problems, or about biology. We'd see a blacktip shark, and he'd say, 'Well, there are this many sharks in the world, and they do this and that.' He was a wonderful adviser. I just learned a huge amount."

Robert MacArthur found a crueler fate. The pain he experienced on a field trip to Arizona in the spring of 1971 proved to signal renal cancer. Kidney removal and chemotherapy came too late, Wilson writes: "The surgeon told him that he had only months or at most one or two years to live. Robert thereafter conducted his life with even greater intensity than before. He completed his final book, *Geographical Ecology.* He journeyed to Arizona, Hawaii, and Panama for more field work, and while at the university"—Princeton now, a full professorship—"he continued to guide his students." His health declined rapidly, beginning in October 1972. Wilson spoke with him at length on the afternoon before he died quietly in his sleep, on 1 November 1972, only forty-two years old.

At the end of the 1960s, Wilson prepared to move on. Having enlarged his field of view from ants to a range of other arthropod species, he had one more major challenge to complete before he shifted from arthropods to vertebrates—from bugs to species with backbones, including his own. All that learning, discovery, collection, and experiment, his and his colleagues', needed to be pulled together, organized, and written down.

7

Full Sweep

As time went on, Ed Wilson took to calling Jim Watson "Caligula"—"the Caligula of biology"—lampooning Watson's empire building and perceived tyranny. Their relationship continued distant and cold. But however much Wilson personally loathed the man's insolence, so different from Wilson's own Southern civility, he never denigrated Watson's achievement. "The deciphering of the DNA molecule with Francis Crick towered over all that the rest of us had achieved and could ever hope to achieve," he would write. "It came like a lightning flash, like knowledge from the gods." Watson and Crick, Wilson felt, "possessed extraordinary brilliance and initiative." The news that Watson had accepted part-time appointment as director of the nearly moribund Cold Spring Harbor Laboratory, on Long Island's North Shore, nevertheless shocked him: "I commented sourly to friends that I wouldn't put him in charge of a lemonade stand." (Later, when Watson successfully boosted both the laboratory's funding and reputation, Wilson would add generously: "He proved me wrong.")

Both men, Watson and Wilson, were moving into the second major phase of their careers. Watson's transformation (after the inevitable years of preparation) appeared to happen almost overnight. In

January and February 1968, *The Atlantic Monthly* serialized *The Double Helix,* Watson's lively memoir of the turmoil, ambition, and sharp elbows of real scientists in pursuit of fundamental discovery and lasting fame; Harvard president Nathan Pusey, scandalized, ordered the Harvard University Press to stop publication of the book scheduled to follow; after which a brash new New York trade publisher, Atheneum, issued it to long best-sellerdom. On 1 February 1968, Watson took up the Cold Spring Harbor directorship. Victor McElheny questioned him at that time, he told me. "The place is falling down," McElheny said to him. "What are you going to do?" Watson said, "We're going to go for cancer and viruses. I have a target of how much money I need for an annual budget, and if I haven't achieved that in *x* time, I will have failed." He set out immediately to raise five million dollars (equivalent to thirty-eight million dollars today) for the laboratory's physical restoration and transformation into a center for cancer research. On 28 March, he married the nineteen-year-old Radcliffe junior Elizabeth Lewis after a yearlong courtship. A week later, on 6 April, he celebrated his fortieth birthday.

His new responsibilities changed him. "Now I'm the head of something," he told *The New York Times* some years later, "and I'm responsible for the cooperation of a large group of people. I have to be responsible. I can't be mercurial anymore." Wilson recalls an occasion in May 1969 when Watson "extended his hand and said, 'Congratulations, Ed, on your election to the National Academy of Sciences.' I replied, 'Thank you very much, Jim.' I was delighted by this act of courtesy." Watson would achieve his goals for Cold Spring Harbor. "In ten years," Wilson comments, "he raised that noted institution to even greater heights by inspiration, fund-raising skills, and the ability to choose and attract the most gifted researchers."

Wilson would transform himself in another direction, not crossing over into administration but enlarging the scale and sweep of his science. Before that transformation, or perhaps as its first stage, a necessary consolidation, he organized into one book everything he and his colleagues had learned about the social invertebrates—wasps, bees, termites, and ants—up to the time of its publication in 1971. *The*

Insect Societies, which the Harvard University Press published in a large coffee-table format, ran to 548 double-column pages clarified with photographs, line drawings, tables, and graphs, and bearing an exhaustive bibliography of 1,701 references, a formidable work of research in its own right. It indexed more papers by Wilson than any other worker—forty-four in all—although he was only forty-two in 1971.

The Insect Societies recalls an equivalent feat in the mid-1930s by Hans Bethe, who would be awarded the Nobel Prize in Physics in 1967. When Bethe arrived in the United States from Germany in 1935 to teach at Cornell University, he found his American colleagues lacking in their knowledge of nuclear physics, then a relatively new field. After teaching several of them the basics, he decided to write down everything he knew about the subject. That work took the form of three articles published in a major physics journal, the first written in collaboration with a colleague, Robert Bacher, the second written alone, the third in collaboration with another colleague. As Bacher recalled, "Bethe wrote these articles seated under a very dim light in Rockefeller Hall at Cornell, a large pile of blank paper on his right and a pile of completed manuscript on his left. Bethe always wrote in ink, with some, but not very many, corrections, even in complicated calculations. He was and is indefatigable, and worked regularly from midmorning till late at night—but would always stop cheerfully to answer questions." The trio of articles came to be called "Bethe's Bible," and served for a decade as the basic textbook in the field.

Similarly, Wilson worked his way through invertebrate biology, filling in missing information as he went along, correcting misjudgments in earlier papers with new information from more recent research, indicating where further research was required, adding his own field observations, cautiously speculating when such speculation seemed warranted. His informed compendium of the life and world of the social insects has never been surpassed.

Ants, along with their cousin wasps and bees, evolved from solitary wasp ancestors during the geologic era known as the Mesozoic, between about 250 million and sixty-five million years ago. (Termites evolved even earlier, from a primitive cockroach.) The Mesozoic was the period when dinosaurs rose to dominance and birds, mammals, and flowering plants first appeared. It ended catastrophically with the massive impact of a ten-kilometer asteroid incoming at forty thousand miles per hour on the coast of what is now the Yucatán Peninsula. The impact splashed molten rock around the world, starting smoky fires that induced a long, dark, cold asteroid winter that starved out the dinosaurs and most other larger life-forms and made room for the evolution of modern plants and animals, including ants and humankind.

In 1967, Wilson and his colleagues Frank Carpenter and Bill Brown reported the discovery of a primitive Mesozoic ant ancestor preserved in a piece of amber found embedded in a beach bluff in Cliffwood, New Jersey. The amber, from a sequoia tree—sequoias grew in New Jersey a hundred million years ago—held two workers of a previously unknown genus and species which the Wilson team named *Sphecomyrma* ("wasp ant") *freyi,* the species designation honoring the amateur mineral collectors who found the amber, a Mr. and Mrs. Edmund Frey.

S. freyi was a curious mosaic of wasp and ant features: short wasplike mandibles; long wasplike antennae; primitive, antlike body; tough, protruding stinger. The three biologists judged it "truly intermediate between the primitive ants and the aculeate [stinging] wasps." It reminded Wilson strongly of *Nothomyrmecia macrops,* the "dawn ant" he had sought unsuccessfully in Australia on his expedition there in 1955. (Bob Taylor, Wilson's former postdoc, finally rediscovered *N. macrops* in 1977, at a site far to the east of where Wilson had searched.) Though the two *S. freyi* specimens appeared as fresh as tomorrow morning, their tomb of sequoia sap had engulfed them and hardened into dark-red amber almost unimaginably long ago. "They are the first undisputed social insect remains of Mesozoic age," the Wilson team concludes, "and extend the existence of social life in insects back to approximately 100 million years."

All modern ants are eusocial, "eu-" meaning in this case "truly": truly social. Wilson lists three traits all eusocial insect species have in common: they cooperate in caring for their young; they divide their reproductive labor, with sterile workers attending fertile queens; and their life cycle is long enough to allow offspring to assist parents.

The first question Wilson asks in *The Insect Societies* is: Why study social insects? He offers scientific answers as well as practical ones. On the scientific side, the social insects in their divisions of labor present for investigation a sort of distributed body. It isn't possible to take apart a living animal, study its various systems, and then reassemble the parts into a living animal again. But it is possible to investigate the collective organism that is a society of wasps, ants, bees, or termites that way—as Wilson says, to study "the full sweep of ascending levels of organization, from molecule to society."

Ecologically, the social insects dominate the land. "In most parts of the earth," Wilson writes, "ants in particular are among the principal predators of other invertebrates. Their colonies, rooted and perennial like woody plants, send out foragers which comb the terrain day and night. Their biomass and energy consumption exceed those of vertebrates in most terrestrial habitats." In the tropics, they move more earth than earthworms do; they're competitive with earthworms even in cold temperate forests.

Charles Darwin, in his delightful last book *The Formation of Vegetable Mould Through the Action of Worms,* estimates the production of new soil by earthworms in one of his fields at an average .083 inch per year, "i.e., nearly one inch in twelve years." Wilson, citing a 1963 study of ant soil turnover in one locality in Massachusetts, a colder climate than England's, reports that ants brought "50 grams of soil to the surface per square yard each year and [added] one inch to the topsoil every 250 years." Earthworms in this comparison moved twenty times as much soil as ants, but earthworms eat their way through the soil and build it up with castings—feces, fluffy and abundant—whereas ants only move soil in the course of building and maintaining their nests. Termites, specializing in dead wood and leaf litter, contribute significantly to soil production as well.

Ants pollinate plants, feed them, and disperse their seeds. Throughout the world, ants protect certain plants, and plants protect ant colonies, in an evolved expression of mutualism. Plants specialized to house ants in various configurations of shelter include bromeliads, coco plums, laurels, legumes, mulberries, orchids, peppers, buckwheats, ferns, madders, coffees, spurges, figs, sapodillas, cacao trees, verbenas, mangoes, sumacs, milkweeds, palms, mahoganies, nutmegs, pitcher plants, figworts, foxgloves, and leatherwoods. Plants feed ants as well, many with specialized food bodies attached to their seeds that spare the seeds themselves for ant distribution. Particularly in poor soils—in grasslands, deserts, and forest margins—ant nests stimulate plant growth. They "turn and aerate the soil," Wilson writes, "add nutrients in the form of excrement and refuse, and hold the ambient temperature and humidity at moderate levels."

So large a group of animals could hardly be only benevolent. Ants are significant pests in some tropical environments and when, like the red imported fire ant, they escape their normal evolutionary setting, where predators and parasites control their population. When it came ashore at young Ed Wilson's doorstep in North America, the fire ant, weedlike, found few competitors to prevent it from exponential increase.

But even in their natural setting, some types of ants are fiercely destructive. Wilson calls the relentless sweep of *Eciton burchelli,* one of several species of army ants of the humid lowland forests of eastern South America, "exciting." It may be for a scientist, but it must be terrifying for any creatures in the path of the swarm that are unable to escape. A "big, conspicuous species," *E. burchelli* army ants form nighttime bivouacs on the march by linking themselves together around their mother queen in chains and nets "that accumulate layer upon interlocking layer until finally the entire worker force"—as many as seven hundred thousand individuals—"comprises a solid cylindrical or ellipsoidal mass up to a meter across," cached in the spaces between the prop roots of the great rain-forest trees. At daylight, the bivouac mass dissolves, the chains and nets break up, and a teeming broil of excited workers tumbles out.

Unlike most army ants, *E. burchelli* is a swarm raider rather than a column raider. At first, Wilson writes, workers fan out in all directions. Then the density of the mass begins to increase "along the path of least resistance and grows away from the bivouac at a rate of up to 20 meters an hour." The increasingly fan-shaped swarm is leaderless, advanced by workers laying odor trails as they press ahead a few centimeters and then wheel back into the mass. Wilson quotes a specialist in army ants, Theodore C. Schneirla, an animal psychologist who was curator of the department of animal behavior at the American Museum of Natural History in New York, describing a typical raid:

> For an *Eciton burchelli* raid nearing the height of its development in swarming, picture a rectangular body of 15 meters [50 feet] or more in width and 1 to 2 meters [3–6 feet] in depth, made up of many tens of thousands of scurrying reddish-black individuals, which as a mass manages to move broadside ahead in a fairly direct path. . . .
>
> The huge sorties . . . bring disaster to practically all animal life that lies in their path and fails to escape. Their normal bag includes tarantulas, scorpions, beetles, roaches, grasshoppers, and the adults and broods of other ants and many forest insects; few evade the dragnet. I have seen snakes, lizards, and nestling birds killed on various occasions; undoubtedly a larger vertebrate which, because of injury or for some other reason, could not run off, would be killed by stinging or asphyxiation.

Unusually in a field report, Schneirla describes the characteristic sounds of an *E. burchelli* raid. There is first of all, he writes, "a kind of foundation noise from the rattling and rustling of leaves and vegetation as the ants seethe along and a screen of agitated small life is flushed out." There's "an irregular staccato" caused by jumping insects knocking against leaves and wood, a sound Schneirla chillingly calls "the collective death rattle of the countless victims." There's "the loud and variable buzzing" of the clouds of flies that hover, circle, or dart immediately ahead of the advancing swarm, much as sea birds do above feeding whales. Flies individually or in small squadrons emit short, higher-pitched notes as they swoop down to capture their share of the escaping prey. Behind this cacophony, antbirds call—any of a large number of species of the Formicariid family, small birds with strong legs and heavy, hooked bills. "One first catches from a distance the beautiful crescendo of the bicolored antbird," Schneirla writes, "then closer to the scene of the action the characteristic low twittering notes of the antwren and other common frequenters of the raid." The antbird seems misnamed, however: it eats not the ants but the prey the ants scare up in their sweeps, not different in that sense from the similarly opportunistic flies.

The question Wilson is asked about ants more than any other, he writes, is whether army ants are "the terror of the jungle." No, he answers, not really:

> Although the driver ant colony is an "animal" weighing in excess of 20 kg [44 pounds] and possessing on the order of 20 million mouths and stings, and is surely the most formidable creation of the insect world, it still does not match up to the lurid stories told about it. After all, the swarm can only cover about a meter of ground every three minutes. Any competent bush mouse, not to mention man or elephant, can step aside and contemplate the whole grass-roots frenzy at leisure, an object less of menace than of strangeness and wonder, the culmination of an evolutionary story as different from that of mammals as it is possible to conceive in the world.

If army ants are relentless raiders, another large ant family has made the transition from hunting to farming. Leafcutter ants of the genera *Acromyrmex* and *Atta,* natives of South and Central America and the Southern United States, cut pieces of leafy vegetation and carry them over their heads like parasols back to their colonies, where they drop them onto the floor of large chambers carved out along a central tunnel. Workers have lined the chambers with the ant equivalent of papier-mâché—chewed-up plant matter—inoculated with a fungus. Smaller workers next clip the dropped leaf pieces into smaller fragments and pass them along to still smaller workers that chew them into moist pellets, anoint the pellets with fecal drops, and plant them in the chambers' pellet garden. Yet smaller workers then clip strands of fungus from the densest areas of growth on the walls and floor of the chamber and pack the strands among the new pellets. "Finally," Wilson concludes his description of this consecutive process, which he compares to an assembly line, "the very smallest and most abundant workers patrol the beds of fungal strands, delicately probing them with their antennae, licking their surfaces, and plucking out spores and strands of alien species of mold." The fungus the ant farmers grow then serves to feed the colony.

The great evolutionary advantage of agriculture over hunting, with ants as with humans, is a more reliable food supply, which can support a larger population. *Atta* colonies are huge, numbering from hundreds of thousands to as many as eight million individual workers. "A full-grown colony," Wilson reports, "consumes approximately the same quantity of plant material as a cow." All the colony's workers are the daughters of a single long-lived queen, whose lifespan may extend from ten to fifteen years. She establishes the colony alone, starting its fungus garden, laying eggs, and tending them until her first workers eclose—emerge as adults from their pupae—and take over all the nascent colony's duties except egg laying, which remains the queen's duty alone. "A rough calculation," Wilson writes, "reveals that the mother of a mature colony lays on average about 20 eggs per minute, thus 28,800 per day and 10,512,000 per year."

Human agriculture originated with the domestication of wild grains between seven and ten thousand years ago. Ant agriculture, which evolved across a period of thirty million years beginning around fifty million years ago, depends on single species of fungus transmitted clonally from colony to colony by founding queens, which carry a small plug of fungus cultivar in a special pouch in their mouths when they leave a colony on their nuptial flights and use it to start their new gardens. These ancient fungus clones may be several million years old.

Between extremes of ant evolution, such as the mass raiding of army ants and the specialized farming of the leafcutters, lie some twelve to twenty thousand species of ants in a great variety of sizes, numbers, and behaviors. There are ants that feed exclusively on bugs, on seeds, on insect honeydew secretions, on other ants. There are ants that live deep underground, ants that live in the tops of trees, ants that tenant hollow twigs, ants that use silk drawn from their larvae to construct tentlike nests, parasitic ants that occupy the nests of other ants. Degenerate slave-making ants that have lost all skills other than raiding depend on their slave colonies for care and survival. In one such species, the worker caste has disappeared entirely, leaving free-living ectoparasite queens, Wilson writes, "modified for riding on the backs of the host queens."

Male ants live short and parasitic lives, fed by workers but performing no worker functions themselves, and scarcely surviving a single summer. Their entire purpose is to inseminate a queen on her initial nuptial flight, when she typically mates with multiple males to collect enough sperm to last her a lifetime. "Flying sperm dispensers," Wilson calls ant males dismissively, recalling the "dancing dons of the cocktail-lounge set" he had sourly imagined were courting Irene when he first glimpsed her descending the stairs at Shirley Hayes.

An unfamiliar aspect of ant organization is what Wilson calls age polyethism (poly-EETH-ism), the changing role of workers as they age. "Young workers tend to remain in the nest and nurse the brood," he reports, "while older workers spend more time outside the nest." Workers aren't specialized or organized into groups to perform specific tasks; they direct their activities to whatever work they find in front of them. So working inside the nest or outside the nest is a tendency rather than a compulsion, and experiments Wilson describes demonstrate that switching brood workers outside leads them to take up foraging, and vice versa.

Even so, young workers usually spend their first weeks caring for the brood, followed by a similar period keeping up the nest—half their time taking care of other workers, handling dead prey, and cleaning the nest, the other half resting. Older workers move outside to perform riskier duties—foraging, patrolling colony territory, and defending the nest against invasion. Unlike human societies, then, ant societies assign their young to the relative safety of home service and send their older members, which are nearer the end of their lives, out onto the land or off to war.

Beyond the natural history of ants lies the more rugged terrain of ant sociology. The central question in that area of knowledge, Wilson proposes, is how complex social behavior emerges from the simple behavior patterns of individual ants. An ant by itself has only a limited repertoire of responses, "neither exceptionally ingenious," Wilson writes, "nor exceptionally complex. The remarkable qualities of [ant] social life are mass phenomena that emerge from the meshing of these simple individual patterns by means of communication."

The meshing is anything but efficient. As Wilson describes it, a collectivity of ants lurching toward a mass action sounds like an episode of the Keystone Kops. "It usually results from conflicting actions of many workers," he explains. If an ant colony "decides" to emigrate from one nest site to another, for example, workers carrying eggs, larvae, and pupae to the new site have to shove past workers returning brood the other way. Other workers "run back and forth carrying nothing." The "decision" to settle in a new nest is made collectively, as a preponderance of individual actions draws in more and more members of the colony. (Bees similarly decide through their behavior; bee swarms can be divided or dispersed when the collectivity fails to coalesce around a new site.)

At this point, Wilson introduces a theory a French entomologist proposed in 1959 that has since come to be applied far beyond the world of insects. Pierre-Paul Grassé was studying termite nest-building behavior. It appeared to him that termites responded to actions performed by other termites—even seemingly random actions—by moving to continue the work. Placed in a container along with bits of mud and excrement, the termites first explored the container. (Termites are blind.) Then they began picking up and putting down pellets of the mud and excrement. If several pellets ended up on top of one another, the termites began adding to the nascent column. If no other columns emerged nearby, they eventually lost interest in the single column they were building. But if two columns happened to be developing near each other, they built up both, and any others as well. As the columns lengthened, the termites began arching them inward toward each other, presumably following an inherited script. As the columns met and were connected, they formed structural arches. Eventually, several such constructions, emerging from random initial actions, produced an intricate, cathedral-like termite nest.

Grassé named this mechanism of coordination "stigmergy," a word he derived from Greek roots meaning "to incite to work." Grassé's definition of stigmergy was limited to the insect behavior he was studying. A more general definition recognizes stigmergy as a feedback loop; as the Belgian cyberneticist Francis Heylighen describes it, "An action produces a mark which in turn incites an action, which produces another mark, and so on." In other words, Heylighen continues, "actions stimulate their own continued execution via the intermediary of the marks they make—where a mark is a perceivable effect or product of an action." Beyond termite nest building, Heylighen cites ants laying pheromone trails—marks that incite other ants to follow while adding their own pheromone marks to the trail. The reverse is also true: when the prey at the end of the trail has been collected, ants returning empty-handed stop their trail marking, the pheromone cloud disperses, and the parade ends.

Wilson in 1971 thought Grassé's model was too simple, but nevertheless saw its importance as a theory of how simple operators could accomplish complex tasks without centralized direction. Drawing on his deep reading, he remembered an early anticipation of the concept of stigmergy in an 1810 book by a Swiss naturalist, Pierre Huber. "From these observations," Wilson quotes Huber, "and a thousand like them, I am convinced that each ant acts independently of its companions. The first that hits upon an easy plan of execution immediately produces the outline of it; others only have to continue along these same lines, guided by an inspection of the first efforts."

Wilson's early references to the phenomenon measure his alert attention to developments in his field. Hardly anyone paid attention to stigmergy until well into the era of computers and programming;

Heylighen found only about one published reference per year from 1960 to 1990, but by 2006 that number had increased to 500, and in 2020 to 10,800. Stigmergy, despite its unappealing name, turns out to be a deep communication structure. Wikipedia is stigmergic, with articles initiated by individuals, then corrected and extended by other individuals, with little or no communication among them. The stock market is stigmergic. So is the changing connectivity of memory neurons in the brain, building and removing connections in response to recall and forgetting. So is the optimization of packet routing in communication networks.

Science itself operates stigmergically: individual or groups of scientists make discoveries, which they publicly communicate in science journals and at conferences. Learning of those discoveries, other scientists use them to pursue related discoveries. Without centralized direction, then—with no central Department of Discovery directing the work—science advances along a broad front marked by what the philosopher of science Michael Polanyi calls "growing points." The principle of this distributed coordination, Polanyi writes, "consists in the adjustment of the efforts of each [scientist] to the hitherto achieved results of the others. We may call this a coordination by mutual adjustment of independent initiatives—of initiatives which are coordinated because each takes into account all the other initiatives operating within the same system."

In a final chapter of *The Insect Societies,* Wilson pointed the way to his next major effort of scientific synthesis. As he had just pulled together everything known up to 1971 about the social organization of invertebrates, he would now attempt to do the same for vertebrates—for the larger and better-known world of animals with backbones, from snakes and birds to primates, including humans. He hadn't planned on doing that work, he told me; he assumed that a specialist in vertebrates would take up the challenge. But when he discussed it with colleagues, no one volunteered. They preferred working within their own specialties to venturing onto the unfamiliar ground of large-scale synthesis.

Wilson might well have rested on his laurels. With tenure at Harvard, with his solid research record, with the publication of *The Insect Societies,* he had established himself as a leading, even *the* leading world specialist on the social insects. Yet several challenges, both personal and professional, pressed him to expand his scientific range.

Personally, he had been thinking about what social behaviors different species had in common since as early as 1956, when he was a new assistant professor at Harvard. As he tells the story in *Naturalist,* he had taken on his first graduate student, Stuart Altmann, a Los Angeles native and UCLA graduate, because Altmann's pioneering interest in studying social behavior among free-living primates had left him without a graduate adviser. Wilson's studies of ant social behavior matched up across the invertebrate-vertebrate divide. "I was hardly more than a graduate student myself," Wilson recalls, "just a year older than Altmann, eager to learn the strange new subject he had chosen."

In June 1956, Altmann had begun studying rhesus macaques on a thirty-seven-acre island preserve, Cayo Santiago, just off the east coast of Puerto Rico. The preserve had been established in 1938 for tropical-medicine research with some five hundred rhesus monkeys that had been captured in India and transported to the preserve by steamship. By 1955, the U.S. National Institutes of Health had purchased Cayo Santiago and its macaque colony for research in behavioral ecology. Altmann's first work, besides rebuilding the decaying docks and housing, was conducting a census and tattooing all the monkeys for identification; he found 225 monkeys in residence at that time.

Wilson joined Altmann for two days of observation on Cayo Santiago in December 1956, an experience Wilson calls "an intellectual turning point":

> When I first stepped ashore I knew almost nothing about macaque societies. . . . As Altmann guided me on walking tours through the rhesus troops, I was riveted by the sophisticated and often brutal world of dominance orders, alliances, kinship bonds, territorial disputes, threats and displays, and unnerving

intrigues. I learned how to read the rank of a male from the way he walked, how to gauge magnitudes of fear, submission, and hostility from facial expression and body posture.

Altmann had warned Wilson to move carefully around the macaque infants or risk attack. If you're challenged, Altmann cautioned, don't stare at the challenger; like most wild animals, macaques interpreted a stare as a threat. Bow your head and look away. Wilson needed the advice. On the second day of his visit, he moved too quickly near a young monkey, which shrieked its distress. "At once the number two male ran up to me," Wilson recalls, "and gave me a hard stare, with his mouth gaping—the rhesus elevated-threat expression. I froze, genuinely afraid. Before Cayo Santiago I had thought of macaques as harmless little monkeys. This individual, with his tensed, massive body rearing up before me, looked for the moment like a small gorilla." Wilson lowered his head and looked away. Eventually, the macaque accepted the gesture of submission and moved off.

Back on mainland Puerto Rico in the evenings—Cayo Santiago is less than a mile offshore—Wilson and Altmann discussed the social behaviors they studied, looking for connections. To Wilson's frustration, they found very few: "Primate troops and social insect colonies seemed to have almost nothing in common." Macaques were organized in dominance orders, each individual known and recognized. Social insects, anonymous and short-lived, flourished in caste-based harmony. In 1956, neither Wilson nor Altmann had the conceptual tools to map out much more than superficial connections, if any. Altmann got busy working on his doctoral dissertation on the Cayo Santiago macaques; Wilson returned to teaching and fighting off the molecular biologists.

Ironically, it was partly Wilson's struggle with the molecular biologists that led him to the approach to a unifying theory that he was looking for. That ongoing challenge had threatened to engulf evolutionary biology and continued to model a more formal and mathematical science. Wilson had titled the final chapter of *The Insect Societies* "The Prospect for a Unified Sociobiology." Though the term "sociobiology" had been used in other contexts as far back as 1912, Wilson appropriated it as a term of art meaning "the systematic study of the biological basis of all social behavior." Sociobiology, he thought, would grow out of population biology as a counterweight to molecular biology, because not everything in biology could be reduced to the molecular level. He had come to believe that "populations follow at least some laws different from those operating at the molecular level, laws that cannot be constructed by any logical progression upward from molecular biology."

To that end, he studied population biology in the late 1960s even as he worked out the observational and theoretical scientific record of the social insects. Not one to waste such an effort, he and his Harvard colleague William H. Bossert, a biologist and applied mathematician, wrote *A Primer of Population Biology,* publishing the densely mathematical book the same year as *The Insect Societies,* 1971.

If none of his colleagues who were vertebrate specialists were prepared to write about the social behavior of the vertebrates, he would have to do that work himself: he was, as he says, a "congenital synthesizer." Formulating a theory of sociobiology, he wrote in the final chapter of *The Insect Societies,* would be "one of the great manageable problems of biology for the next twenty or thirty years." As it turned out, it only took him four.

8

Ambivalences

Ed Wilson recalls encountering "the most important idea of all" for sociobiology, a key breakthrough, in the summer of 1964, when he opened his briefcase to catch up on his reading during one of his regular train trips between Boston and Miami. He traveled to Florida by train for his research there because he had promised Irene to avoid air travel until their daughter reached high-school age. "I found an advantage to the restriction," he remembers. "It gave me . . . eighteen hours in a private roomette, trapped by my pledge like a Cistercian monk with little to do but read, think, and write." He wrote his chapters of *The Theory of Island Biogeography* on such travels.

This time, scanning the July 1964 issue of the *Journal of Theoretical Biology,* a three-year-old journal published at that time in London, he encountered the paper he would share with his early graduate student, Dan Simberloff: "The Genetical Evolution of Social Behaviour," W. D. Hamilton's paper on kin selection. He was impatient with it at first. "I was anxious to get the gist of the argument and move on to something else, something more familiar and congenial. The prose was convoluted and the full-dress mathematical treatment difficult, but I understood his main point . . . quickly enough."

Hamilton's main point concerned Darwin's old bête noire, altruism. If natural selection occurs only at the level of individuals, as Darwin believed, then how is it possible for some individuals—a worker bee or ant, a prairie dog barking to alert its town at the approach of a predator—to put themselves at risk or even forgo reproduction in support of their society? Shouldn't selection have snuffed out such unselfish but self-destructive behavior long ago? Darwin had to evoke group selection to explain altruism, but he had never shown convincingly how group selection could work.

The breakthrough answer Hamilton proposed in his 1964 paper was that the unit of selection wasn't in fact the individual, much less the group; it was the gene. The effect is most pronounced in bees and ants, which reproduce according to a system that biologists call "haplodiploidy." (The word is formed from Greek roots, *haplo-* meaning "single," *diplo-* meaning "double.") In normal sexual reproduction, males and females each contribute a complete set of chromosomes to their offspring. In haplodiploid species, however, males develop from unfertilized eggs. An unfertilized egg carries only one set of chromosomes—the female's. When the resulting male offspring then inseminates a female in its turn, it delivers only one set of chromosomes to the union, whereas the female delivers two sets. The product of that mating is then a female with 75 percent female-derived chromosomes and only 25 percent male-derived chromosomes instead of the normal 50–50. That means a worker bee or ant is a super-sister; only 25 percent of its genes are of male descent, 75 percent of female descent. So supporting its mother queen and her other offspring—its super-sisters—passes along more of its genes—75 percent (75–25)—than if the worker itself reproduced—only 50 percent (50–50).

In haplodiploid species, the altruism is one for one. In species that reproduce without such special genetic conditions, altruism as Hamilton frames it involves larger numbers of relatives. A prairie dog, losing its life to a predator after alerting its populous town to the danger, spares the genes it shares with its blood relatives. Or, as Hamilton wrote in his 1964 paper, "We expect to find that no one is prepared to sacrifice his life for any single person but that everyone will sacrifice it when he can thereby save more than two brothers, or four half-brothers, or eight first cousins." (Hamilton's formulation doesn't exclude other altruistic relationships; a husband saving his wife from drowning, for example, potentially protects his genetic investment in their existing or potential offspring, even though they themselves are not blood relatives.)

Wilson's first response to Hamilton's revelation, he recalls, was to dismiss it out of hand. "Impossible, I thought; this can't be right. Too simple. He must not know much about social insects." Some deeper instinct made Wilson hesitate, however, and after he switched over to his Florida train at New York's Pennsylvania Station, he continued assessing Hamilton's paper:

> As we departed southward across the New Jersey marshes, I went through the article again, more carefully this time, looking for the fatal flaw I believed must be there. . . . Surely I knew enough to come up with something. . . . By dinnertime, as the train rumbled on into Virginia, I was growing frustrated and angry. Hamilton, whoever he was, could not have cut the Gordian knot. Anyway, there was no Gordian knot in the first place, was there? I had thought there was probably just a lot of accidental evolution and wonderful natural history. And because I modestly thought of myself as the world authority on social insects, I also thought it unlikely that anyone else could explain their origin, certainly not in one clean stroke.

The next morning, he was still frustrated, but by midday, when the train arrived in Miami, he had found his way through. "I was a convert, and put myself in Hamilton's hands."

Who was this paragon, W. D. Hamilton? Why had Wilson not heard of him before? That fall, he sailed to London on the *Queen Mary* to deliver an invited lecture on the social insects before the Royal Entomological Society. Without benefit of Hamilton's breakthrough, Wilson had published a paper, "The Social Biology of Ants," in the 1963 edition of the *Annual Review of Entomology*, which had prompted the invitation. Hamilton was attending as well, and the day before giving his lecture, Wilson looked him up.

Hamilton was, it turned out, twenty-eight years old and still a graduate student. This major revision of Darwin, a paradigm shift, was the work of a graduate student, whom Wilson found to be "in some respects the typical British academic of the 1950s—thin, shock-haired, soft-voiced, and a bit unworldly in his throttled-down discursive speech." Something of Hamilton's latent force might have been signaled by the missing first joints of his left hand, which Wilson would learn had been mangled during World War II, when young Hamilton had tried to make a bomb in his engineer father's basement laboratory at their home in Kent and succeeded all too well. (His father was making bombs for the Home Guard in those days, against the possibility of a German invasion.)

Walking London, the two men explored their common ground. Hamilton told Wilson of his troubles finding a Ph.D. adviser who credited his work on kin selection. "I thought I understood why," Wilson quips. "His sponsors had not yet suffered through their paradigm shift." They also had been unprepared for Hamilton's formidable mathematical treatment of a subject more usually approached through natural history, descriptively.

Hamilton's insight had been no overnight inspiration. It followed three hard years of work, years when he had struggled not only to find encouragement and support but also to quiet his own doubts. "Make it seem inevitable," Louis Pasteur supposedly advised his students when they prepared to write up their research. Sometimes discovery actually happens that way. More often it needs anxious, multiple iterations to grow from first uncertain hunch to full understanding.

"One long, wild swing of an emotional pendulum" is how Hamilton remembers the painful years from 1960 to 1964, when he was developing his ideas. "Through most of [that period], my feeling of alienation in my work continued. At times I was sure I saw something that others had not seen. . . . At others I felt equally certain that I must be a crank. How could it be that respected academics around me, and many manifestly clever contemporary graduate students that I talked to, would not see the interest of studying altruism along my lines, unless it were true that my enterprise were bogus in some way that was obvious to all of them but not to me?" Lodging in a cheap London bed-sitting room, passing like a ghost unnoticed through the halls of the laboratory where he had study privileges, whose director suspected him of concocting loathsome eugenic theories, he was intensely lonely. "Rather than return to my room," he recalls of the hours after the library where he studied closed, "I would go to Waterloo Station, where I continued reading or trying to write out a model sitting on the benches among waiting passengers in the main hall." He didn't want to talk to them, but he welcomed the human warmth of their presence.

By the end of 1962, with a brief synopsis of his theory accepted for publication in *The American Naturalist,* Hamilton decided he "wasn't a crank." At the same time, he was "utterly tired" of the longer and more generalized paper he'd been writing. As soon as he submitted it to *The Journal of Theoretical Biology,* he wrote to a distinguished Brazilian entomologist, Warwick Estevam Kerr, asking if he might work with Kerr at the University of Rio Claro in São Paulo. Hamilton had realized, he writes, "that I had somehow by-passed (or as I said tunneled under) an essential stage of scientific development. I ought now to make up, to fact-pile something somewhere, test something, become respectable." What Wilson had done at the outset of his career with his extended collecting expedition through the South Pacific, Hamilton would make up with a yearlong stay in Brazil, collecting insects in the Brazilian rain forest and discussing his ideas with Kerr. He would need the time: the editor of *Theoretical Biology,* he says, found his paper "generally acceptable to the journal but [in need of] major revision, and in particular [it] must be split into two." The voyage from London to Brazil, on a freighter, took a month. Learning Portuguese, settling in at Kerr's lab, and starting practical research further slowed the work of revision.

The revised two-part paper went back to London early in 1964 and was accepted for publication. It came out in July, Hamilton reminisces:

> I was traveling up mainly overland from São Paulo towards Canada on my way home to Britain. Very probably the sun of the day that witnessed my paper going into the post . . . would have seen me weaving my old American jeep between the corrugations, stones, and potholes of the Belém-Brasilia road. . . . At midday it would have blazed near vertically on the top of my head as I stopped at the roadside and collected wasps from some nest; later at sunset, if still able to pierce the haze, it would have seen me and my Brazilian companion, Sebastião Laroca . . . slinging our hammocks between low cerrado trees not far back from the stony or sandy piste where occasional lorries still groaned on into the night. For sure, both that day and that night, I was blissfully untroubled about the finer points of measuring relatedness.

After their London walkabout, Wilson decided to feature this unknown graduate student's work in his lecture the next day before the assembled leaders of British entomology. He spent a third of his lecture presenting Hamilton's ideas. "I expected opposition," he writes mischievously, "and, having run through the gamut of protests and responses in my own mind [on his summer train ride], I had a very good idea of what the objections would be. I was not disappointed. . . . When once or twice I felt uncertain [of an answer] I threw the question to young Hamilton, who was seated in the audience. Together we carried the day."

Bill Hamilton took his Ph.D. in 1968, offering two other papers and an introductory essay along with those of 1963 and 1964 in lieu of a separate dissertation. Then, with his new wife, Christine, a London dentist and the prize of a year's courtship, he went off to Brazil's vast Mato Grosso region on a nine-month expedition cosponsored by the Royal Society and the National Geographic Society. A full career of original work followed, along with lecturing at University College London, a professorship in evolutionary biology at the University of Michigan, and, from 1980 until his death in 2000, a research professorship at Oxford.

For Wilson, kin selection was the fulcrum he needed on which to lever the new science he had named sociobiology—"a discipline," he clarifies, "not a particular theory." If social behavior correlates with genetic relatedness, then evolution intrudes into activities long believed to be purely learned from experience, not inherited. By 1971, as Wilson published *The Insect Societies,* with its final chapter pointing the direction, he was ready to begin researching and writing the path-breaking work he would title *Sociobiology: The New Synthesis.*

In 2019, Wilson told an interviewer that the success of the insect book—it was a finalist for a 1972 National Book Award, which surprised him—"led me to thinking I should next do a similar review of vertebrates—mammals, reptiles, amphibians, fishes." Yet *Sociobiology* would not be only the vertebrate counterpart to *The Insect Societies.* It would be a far more ambitious project—"the attempt," a reviewer would later characterize it, "to produce and legitimize a new scientific discipline." The final chapter of Wilson's insect book opens with a bold statement of his next goal, one that alludes to his "intellectual turning point" with Stuart Altmann on Cayo Santiago: "When the same parameters and quantitative theory are used to analyze both termite colonies and troops of rhesus macaques, we will have a unified science of sociobiology." Such was the challenge Wilson would now undertake—"roused," as he admits, "by the amphetamine of ambition":

> Go ahead, I told myself, pull out all the stops. Organize *all* of sociobiology on the principles of population biology. I knew I was sentencing myself to a great deal more hard work. *The Insect Societies* had just consumed eighteen months. When added to my responsibilities at Harvard and ongoing research program in ant biology, the writing had pushed my work load up to eighty-hour weeks. Now I [would invest] two more years, 1972 to 1974, in the equally punishing and still more massive new book.

If that work sounds grim, Wilson counters with a surprise. "In fact," he says, "the years spent writing the two syntheses were among the happiest of my life." They were happy because his family life was happy. He had explored setting up a research station in Florida where Irene and Cathy could join him during his research seasons. That plan had not worked out, but a better one had replaced it.

In 1969, Larry Slobodkin invited Wilson to join him in teaching a summer ecology course at the Marine Biology Laboratory in Woods Hole, Massachusetts, on Cape Cod. The MBL provided cottages for its summer faculty; the Wilsons' stood a mile down the road from the old white Nobska Light, a forty-foot brick-lined iron lighthouse first lit in 1876, which was still sounding its fog signal twice every thirty seconds almost a century later. New Bedford beckoned across Buzzards Bay to the west, Martha's Vineyard across Nantucket Sound to the east, the two channels sparkling with sailboats on sunny days. "Cathy," Wilson recalls, "just then entering kindergarten age, fell in with a gang of other faculty youngsters. She and I also spent hours gazing at butterflies, birds, and, in the swamp behind our cottage, a colony of muskrats." In the early evenings, the family sometimes explored the southern Cape by car. Entering his forties, Wilson began running for weight control and took long afternoon runs along the coastal road down to Falmouth and back. The rest of his free time, he wrote, and read, and wrote. "We continued to return to Woods Hole for another eighteen summers," he reminisces, "through Cathy's college years. It was a balanced life during that long period, deeply fulfilling."

In this comfortable setting, in the summer of 1972, Wilson began researching and writing *Sociobiology,* his most ambitious book. He drew from an even greater range of references than that of *The Insect Societies*—2,552 in all, compared with 1,701 for the earlier book.

Technical books can be difficult to begin; the author either has to assume that the reader already understands the basics of his subject or must somehow bring the reader up to speed. Wilson clearly didn't expect his readership to be prepared already for a new, challenging, and cross-disciplinary science. *Sociobiology* begins with a brief opening chapter, followed by a chapter of "Elementary Concepts," followed by a chapter on "The Prime Movers of Social Evolution," followed by a chapter on "The Relevant Principles of Population Biology." Only then, at page 106, does Wilson begin his discussion of sociobiology itself, with a chapter titled "Group Selection and Altruism."

The brief opening chapter, "The Morality of the Gene," is not the usual formal précis. To the contrary, it's profoundly personal. The personal is partly screened behind indirect language, but once you pierce the screen, it's startling in its nakedness. The first sentence in a book of almost seven hundred pages, composed with the grand program of founding a new biological science, is "Camus said that the only serious philosophical question is suicide."

Camus is Albert Camus, the Algerian-born French journalist, novelist, playwright, and World War II resistance fighter whose writing contributed to the postwar development of the philosophy known as existentialism. Camus won the Nobel Prize in Literature in 1957 at only forty-four years of age. Two years later, he was killed in a car accident, a figure of enduring glamour and mystery. Wilson, after quoting this world celebrity in the first sentence of his book, immediately challenges him. The only serious philosophical question is suicide? "That is wrong even in the strict sense intended," his second sentence disputes.

Then, as if abashed by his own boldness, Wilson retreats to third-person indirection, stepping back from the challenge of his beginning. This child of a suicide—and who knows better how serious an act a suicide might be?—becomes "the biologist," both distancing himself from the subject and preparing to visit his scorn upon philosophers who presume to speak of profound human experience without a matching depth of science:

> The biologist, who is concerned with questions of physiology and evolutionary history, realizes that self-knowledge is constrained and shaped by the emotional control centers in the hypothalamus and limbic system of the brain. These centers flood our consciousness with all the emotions—hate, love, guilt, fear, and others—that are consulted by ethical philosophers who wish to intuit the standards of good and evil.

The pejorative word here is "intuit." For Wilson, to merely "intuit" standards of good and evil is hardly better than guesswork. If self-knowledge is constrained by the emotional control centers, what constrains the emotional control centers?

> What, we are then compelled to ask, made the hypothalamus and limbic system? They evolved by natural selection. That simple biological statement must be pursued to explain ethics and ethical philosophers, if not epistemology and epistemologists, at all depths. Self-existence, or the suicide that terminates it, is not the central question of philosophy. The hypothalamic-limbic complex automatically denies such logical reduction by countering it with feelings of guilt and altruism. In this one way the philosopher's own emotional control centers are wiser than his solipsist consciousness, "knowing" that in evolutionary time the individual organism counts for almost nothing.

The density of these opening paragraphs reflects the density of expectation they incorporate: Wilson's ambition as a biologist to supersede the intuitions of traditional philosophy and found a better science; his growing confidence that a sociobiological approach might bear fruit; and, privately, his desire as a son burdened with "feelings of guilt and altruism," to understand and contextualize his father's act now that he might finally have the tools to do so.

By implication, kin selection played a part in Wilson's father's decision to take his own life. When Ed Senior altruistically—self-destructively—released his son from the guilty burden of supporting him and Pearl, he allowed Ed Junior to advance the family's achievement, to move up in the world socially, culturally, and financially, carrying Ed Senior's genes with him.

Returning to Camus, Wilson corrects the French philosopher's celebrated conclusion: "To his own question, 'Does the Absurd dictate death?' Camus replied that the struggle toward the heights is itself enough to fill a man's heart." Thus the myth of Sisyphus, which

Camus explores, the deceitful king of Corinth whom Zeus condemned forever to roll a great stone up a hill in Hades. "But every time," Odysseus recounts in the *Odyssey,* "as he was about to send it toppling over the crest, its sheer weight turned it back, and once again towards the plain the pitiless rock rolled down." Whereupon Sisyphus descends to begin his absurd labor again.

Wilson calls Camus's conclusion that Sisyphus is fulfilled "arid" but "probably correct," adding, however, that "it makes little sense except when closely examined in the light of evolutionary theory." And what that light reveals is:

> The hypothalamic-limbic complex of a highly social species, such as man . . . has been programmed to perform as if it knows . . . that its underlying genes will be proliferated maximally only if it orchestrates behavioral responses that bring into play an efficient mixture of personal survival, reproduction, and altruism. Consequently, the centers of the complex tax the conscious mind with ambivalences whenever the organisms encounter stressful situations. Love joins hate; aggression, fear; expansiveness, withdrawal; and so on, in blends designed not to promote the happiness and survival of the individual, but to favor the maximum transmission of the controlling genes.

This explication of Camus's program may explain the absurdity of existence intellectually; it hardly comprehends that absurdity emotionally. The emotions of ambivalence Wilson is feeling burst forth in the next paragraph, not only the biologist speaking now but also the man—and the son. The "genetic consequences" of these ambivalences he will explore formally "later in this book":

> For the moment, suffice it to note that what is good for the individual can be destructive to the family; what preserves the family can be harsh on both the individual and the tribe to which the family belongs; what promotes the tribe can weaken the family

> and destroy the individual, and so on upward through the permutations of levels of organization.

If it hasn't been obvious before, here is vulnerable, direct expression of Wilson's deeply tragic view of life.

He will explore it further later; for now, having found a way to understand his father's abandonment as a gift rather than wholly a betrayal, he expansively, as if released, lays larger claim to humanism's territory. Sociology, he writes, is not yet sociobiology; it attempts to explain human behavior by "empirical description" but (like philosophy) by "unaided intuition," without reference to evolutionary explanation. Intuition isn't science, however, and sociology, Wilson goes on, falls far short of the level of generalization and testable prediction that real science demands. He invokes the Modern Synthesis of evolutionary theory—Darwin reinforced and corrected by Mendel—to emphasize what has been possible for other previously "intuitive" sciences such as taxonomy and ecology. And then, with breathtaking bravura, he sweeps all the social disciplines into his net, conceding honestly as he does so that he doesn't yet know if it's strong enough to hold them:

> It may not be too much to say that sociology and the other social sciences, as well as the humanities, are the last branches of biology waiting to be included in the Modern Synthesis. One of the functions of sociobiology, then, is to reformulate the foundations of the social sciences in a way that draws these subjects into the Modern Synthesis. Whether the social sciences can be truly biologicized in this fashion remains to be seen.

Wilson's final claim for his new science addresses the challenge that motivated him to seek to expand and modernize evolutionary biology in the first place—the challenge of the new disciplines he and his colleagues had been competing with in their war for survival: "I believe that [sociobiology] has an adequate richness of detail and aggregate of self-sufficient concepts," he writes, "to be ranked as coordinate with such disciplines as molecular biology and developmental biology."

"Pull out all the stops," indeed. It was a mighty pipe organ he was warming up to play.

Sociobiology: The New Synthesis is a daunting book to read, at 697 pages of wide double columns, weighty with technical terms—weighty literally at more than five pounds. Nor is it a work easily summarized. As with *The Insect Societies,* but even more comprehensively, Wilson ranges across the entire literature of animal life up to the early 1970s, looking for evidence of underlying patterns and connections, reporting them when he finds them, making connections himself, and noting lines of potential further research. Where his findings connect back to the invertebrates, as they often do, he brings in those linkages and parallels as well. This immense work of synthesis is comparable to Mendeleev's work in the second half of the nineteenth century, assembling the periodic table of the elements, which not only revealed the underlying relationships among the chemical elements but also, by opening space for missing elements, led in time to their identification.

There are an estimated 8.7 million known species of plants and animals on Earth, plus or minus 1.3 million, of which 2.2 million are marine. Another estimated 86 percent of land species and 91 percent of ocean species remain to be described. The time would come when Wilson would organize a worldwide effort to remedy this neglect; for now, he was concerned to find the social commonalities among identified vertebrate species.

Early on in *Sociobiology,* Wilson tackles the seemingly intractable problem of classification. "All previous attempts to classify animal societies have failed," he flatly declares. Defining societies according to their traits has evidently led to a biological version of the universal conflict between lumpers and splitters: lumpers keep it simple; splitters compile long lists. Wilson's predecessor William Morton Wheeler started with only three categories of social traits—active versus passive; primarily reproductive, nutritive, or defensive; colonial or free-ranging—and from them generated five basic kinds of animal societies. At the other extreme, the early twentieth-century German zoologist Paul Deegener identified no fewer than forty categories of traits. Wilson's impatient conclusion: "Classification based upon all relevant traits is a bottomless pit." The only way to avoid it, he writes, is to turn to the social qualities themselves—specific features rather than general characteristics—and catalogue them.

Wilson identifies ten "qualities of sociality" as a starting point, all of them measurable. They range from group size, to how populations are distributed, to how close group members are to one another, to how open or closed a society is to immigration, to how specialized the roles are that its members assume, to how closely members work together (recall those pheromone trails), to information flow, to how much time members devote to social behavior (lemurs, he writes, an only marginally social species, about 10 percent; macaques, 80 to 90 percent).

These ten qualities serve as Wilson's tool kit for his extended exploration of vertebrate sociobiology. For a review at the full depth of his findings you should study the book itself; here I can only offer a few examples to show how Wilson works.

Less than thirty pages along in *Sociobiology,* he emphasizes the importance of developing *testable* theory—which suggests how much he judged needed to be done at that point in time to advance evolutionary biology from natural history to real science. "The goal of investigation," he writes, "should not be to advocate the simplest explanation but rather to enumerate all of the possible explanations, and then to devise tests to eliminate some of them." Not "quick and elegant laboratory experiments," he adds, implicitly sideswiping the molecular biologists, but long, hard fieldwork.

For one example of what he means, he cites an analysis of the courtship displays of the goldeneye duck in which researchers recorded twenty-two thousand feet of film of field observations, compiling from it a list of recorded displays, and measuring the duration of each display.

The German field biologist George Schaller, Wilson continues his examples, in a three-and-a-half-year study of Serengeti lions, "spent 2,900 hours and traveled 149,000 kilometers while locating and monitoring several prides on a daily basis." Two Japanese scientists studying the changes that occur in the labor programs of honeybee workers as they age "spent 720 hours collecting data on 2,700 individually marked bees." Another bee investigator "watched a single worker for a total of 178 hours and 45 minutes." Wilson's admiration for such yeoman fieldwork shines through his careful enumerations. He concludes, "The data yielded by such efforts become clinical in detail: each individual can be recognized, its idiosyncrasies recorded, and the development of its social status to some extent charted. Then the fine structure of the communication network begins to emerge. . . . This new level of information is vital to the future development of sociobiology."

Some environmental factors tend to induce social evolution, Wilson observes a little later; others do not. The most frequently reported factor, he writes, is defense against predators. Crowding together—Wilson calls it "centripetal movement"—and pushing to the middle of the resulting crowd tends to protect the stronger animals but to put the weaker animals at the edge of the crowd at greater risk; Bill Hamilton, Wilson comments, called this action the "selfish herd" strategy. The results are visually impressive but minimally organized: "Centripetal movement generates not only herds of cattle but also fish and squid schools, bird flocks, heronries, gulleries, terneries, locust swarms, and many other kinds of elementary motion groups and nesting associations."

More familiar to many observers are strategies that swamp predators with so much prey that they're unable to consume more than a small part of the whole. Few with access to television have failed to see footage of green turtle hatchlings emerging simultaneously in large numbers on a beach and dashing desperately en masse to the ocean, dodging hungry gulls. Even more remarkable are the mass emergences of periodical cicadas at thirteen- and seventeen-year intervals in the Eastern United States. These big, harmless, noisily buzzing insects—seven different species are known—spend long years underground as vegetarian nymphs, only to emerge simultaneously in vast numbers, crawl halfway up the trunks of trees, molt their nymphal skins, shake out their wings, and fly off for a month or two of mating and egg laying before they perish. "No ordinary predator species," Wilson writes, "can hope to adapt specifically to a prey that gluts it for a few days or weeks and then disappears for years." He quotes a 1962 study by two entomologists, R. D. Alexander and T. E. Moore, describing the phenomenon:

> In some years practically all of the population in a given forest emerges on the same night. . . . In 1957, Alexander witnessed such an emergence in Clinton County, Ohio. In a woods that during the afternoon had contained only scattered nymphal skins and no singing individuals . . . nymphs began to emerge in such tremendous numbers just past dusk that the noise of their progress through the oak leaf litter was the dominant sound across the forest. Thousands of individuals simultaneously ascended the trunk of each large tree in the area, and the next morning foliage everywhere was covered with newly molted adults. . . . In this case, it was literally true that the periodical cicadas had emerged as adults within a few hours from eggs laid across a period of several weeks seventeen years before.

A contrasting group behavior, Wilson writes, is acting together socially as predators. Lion prides and packs of wild dogs or hyenas are obvious African examples. Killer whales work together to hunt porpoises. "The killer whales encircled the porpoises," Wilson describes a large hunt reported by two researchers, "then gradually constricted the circle to crowd the porpoises inward. Suddenly one whale charged into the porpoises and ate several of the trapped animals while its companions held the line. Then it traded places with another whale, who fed for a while. The procedure continued until all the porpoises were consumed." (Humpback whales visit the bay on the Pacific Coast where I live and similarly ball up schools of sardines migrating north in early summer, then open their mouths and swim up through the sardine ball, breaching almost entirely out of the water with full mouths as a cloud of screaming gulls rains down to feed on the spill.)

Other group strategies for maximizing fitness are less obvious. Wilson mentions excluding strangers from a population's territory—an obvious way to stabilize a population—and modifying the environment. As a sophisticated example of environmental modification, he describes the extraordinary mound nests of an African fungus-growing termite, *Macrotermes bellicosus:*

> As air warms in the central core of the nest from the metabolic heat of the huge colony inside, it rises by convection to the large upper chamber and then out to a flat, capillary-like network of chambers next to the outer nest wall. In the outer chambers, the air is cooled and refreshed [by radiation and diffusion through the thin walls]. As this occurs, it sinks to the lower passages of the nest [again].

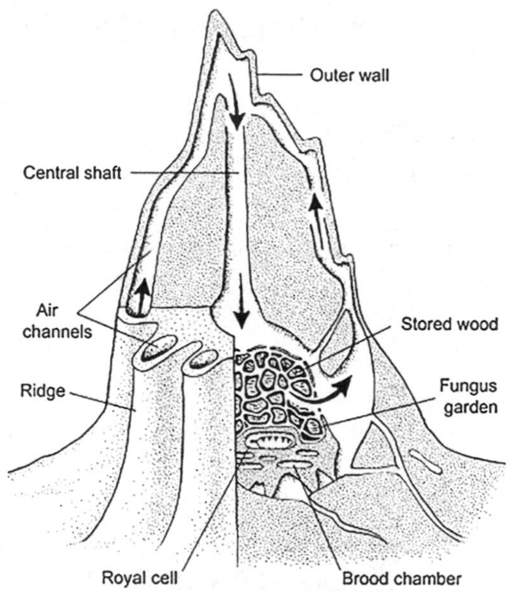

African fungus-growing termites cool and freshen their nests with a convection-driven air circulation system.

Sociobiology is rich in stories, examples, and occasional flashes of humor that lighten its work of hard science. Here is Wilson tweaking culture snobs by elevating the value of fishing, boxing, and football:

> It is a fact worrisome to moralists that Americans and other culturally advanced peoples continue to devote large amounts of their time to coarse forms of entertainment. They delight in mounting giant inedible fish on their living room walls, idolize boxing champions, and sometimes attain ecstasy at football games. Such behavior is probably not decadent. It could be as psychologically needed and genetically adaptive as work and sexual reproduction, and may even stem from the same emotional processes that impel our highest impulses toward scientific, literary, and artistic creation.

Little or nothing in the first twenty-six chapters of this big, twenty-seven-chapter book seems to have raised more than technical objections among its scholarly and scientific readers. The reception of chapter 27, however, "Man: From Sociobiology to Sociology," started an ugly and painful culture war that changed the direction of Wilson's life.

9

Human Natures

When Ed Wilson began writing *Sociobiology,* he told me, he "hadn't yet considered including humans, but when I got going, I was soon to see that I needed to do so." A synthesis that included all the rest of the vertebrate kingdom but left out a species that considers itself the kingdom's crowning glory would be derelict indeed. At the same time, Wilson understood that he would be treading on sacred ground. He would make only glancing reference to *Homo sapiens* in earlier chapters of the book; even in chapter 27, his chapter on human sociobiology, his approach would be circumspect.

At the beginning of chapter 27, Wilson invites the reader "to consider man in the free spirit of natural history, as though we were zoologists from another planet completing a catalog of social species on Earth." He appeals, that is, for a long view and an open mind. We're "a very peculiar species," he begins, with the widest geographical range and the highest local density of any of the primates. We've "preempted all the conceivable hominid niches," leaving us the only species of *Homo* still around.

We're peculiar in another regard, "anatomically unique": upright posture, nearly hairless, many more sweat glands than our primate cousins—all the makings of a daytime hunter in our East African Eden, capable of running our animal prey to overheated exhaustion, with powerful thumbs for grasping spears to finish it off.

Biologically, we've shed the compulsion of estrus—female heat, signaled in other mammals by periodic genital color changes and the release of pheromones—in favor of nearly continuous sexual availability. "The flattened sexual cycle and continuous female attractiveness cement the close marriage bonds that are basic to human social life."

Most peculiar of all is our large, globular head, which "a perceptive Martian zoologist would regard . . . as a most significant clue to human biology":

> The cerebrum of *Homo* was expanded enormously during a relatively short span of evolutionary time. . . . Three million years ago [our ancestor] *Australopithecus* had an adult cranial capacity of 400–500 cubic centimeters, comparable to that of the chimpanzee and gorilla. [That volume increased to] 900–2000 cubic centimeters in modern *Homo sapiens*. The growth in intelligence that accompanied this enlargement was so great that it cannot yet be measured in any meaningful way.

Wilson calls this unique enlargement of the brain "mental hypertrophy." We've surged forward in mental evolution, he notes, with results that almost defy analysis. "Individual species of Old World monkeys and apes have notably plastic social organizations; man has extended the trend into a protean ethnicity." Monkeys and apes scale the intensity of their social and sexual interactions; in humankind such accommodations are "multidimensional, culturally adjustable, and almost endlessly subtle." Bonding and reciprocal altruism are rudimentary in other primates; "man has expanded them into great networks where individuals consciously alter roles from hour to hour as if changing masks."

One goal of human sociobiology, then, is to trace these and other qualities back through time, to determine to what extent they're vestiges of evolutionary development and to what extent adaptations to modern cultural life. To distinguish between these two qualities, known historically as "nature" and "nurture," Wilson borrows a term from sociology, the "biogram." The biogram is everything we've inherited in the course of our evolution: loosely, our nature. "Our civilizations," Wilson writes, "were jerrybuilt around the biogram." How have they been influenced by it, he asks, and how have they influenced it in turn? The work of human sociobiology, the capstone of Wilson's new science, is to find out.

Nor is that work likely to be easy: "Experience with other animals indicates that when organs are hypertrophied, phylogeny [i.e., a species' history of evolutionary descent] is hard to reconstruct. This is the crux of the problem of the evolutionary analysis of human behavior." The crux of the problem, that is, is to sort out which human behaviors evolved by natural selection and which we learned from experience, and to do so amid the tumult of an evolutionary history extravagant with several hundred thousand years of human culture.

The conclusion of Wilson's chapter 27 could be only the merest sketch of what he imagined human sociobiology might become. He discusses those expectations in the final section—"The Future"—of this final chapter of his long, comprehensive book. In the future, Wilson writes, "when mankind has achieved an ecological steady state, probably by the end of the twenty-first century," biology should be at its scientific peak and the social sciences "rapidly maturing." He was writing in 1973, which meant he was allowing about 125 years for the biological sciences to master human biology to this deep level:

> The transition . . . to fundamental theory in sociology must await a full neuronal explanation of the human brain. Only when the machinery can be torn down on paper at the level of the cell and put together again will the properties of emotion and ethical judgment come clear. Simulations can then be employed to estimate the full range of behavioral responses and the precision of their homeostatic controls. . . . Cognition will be translated into circuitry. Learning and creativeness will be defined as the altera-

> tion of specific portions of the cognitive machinery regulated by input from the emotive centers. Having cannibalized psychology, the new neurobiology will yield an enduring set of first principles for sociology.

Wilson's use in this description of mechanical metaphors—"circuitry" and "machinery" and "input"—is unusual for him, a borrowing from information theory. He continues in that vein: "[Sociobiology] will attempt to reconstruct the history of the machinery and to identify the adaptive significance of each of its functions." Some functions, he predicts, "are almost certainly obsolete," those in particular that derive from our days of hunting, gathering, and intertribal warfare. Others—and here he paraphrases his discussion in the opening pages of the book—"may prove currently adaptive at the level of the individual and the family but maladaptive at the level of the group—or the reverse." What was tragic in the past will be tragic in the future.

With that echo reverberating, Wilson alludes to another future, the utopian (or, arguably, dystopian) vision of his Harvard colleague B. F. Skinner, the radical behaviorist. Skinner's vision invokes the historically compromised subject of eugenics, the discipline of deliberate human "improvement." As Wilson assesses it:

> If the decision is taken to mold cultures to fit the requirements of the ecological steady state, some behaviors can be altered experientially without emotional damage or loss in creativity. Others cannot. Uncertainty in this matter means that Skinner's dream of a culture predesigned for happiness will surely have to wait for the new neurobiology. A genetically accurate and hence completely fair code of ethics must also wait.

Wilson judges that the creation of such a "planned society . . . seems inevitable in the coming century." Inevitable it may be, he writes, but it will be no utopia, because the functional and the dysfunctional—the good and the bad—are often genetically linked.

A planned society that ablates the dysfunctional, whatever that might be—war, hostility to outsiders—must inevitably ablate some part of the functional as well. "In this, the ultimate genetic sense," Wilson concludes, "social control would rob man of his humanity."

Wilson thus finds tragedy in the collision of two forces—the inexorable drive of evolution against humanity's desire to control its environment, including its own dysfunctions—nature versus nurture at a far more conflictive extreme.

As he turned to Albert Camus at the beginning of *Sociobiology* for Camus's explication of personal tragedy, so Wilson turns to the French writer again at the end, quoting Camus's "foreboding insight" into the continuing scaling away of human aspiration. The old verities that sustained humankind, at least in the West, had been the Earth as the center of the universe, man as a separate creation, and human beings as creatures of reason. Copernicus had demolished the first, Darwin the second, and Freud the third. As Camus concludes:

> A world that can be explained even with bad reasons is a familiar world. But, on the other hand, in a universe suddenly divested of illusions and lights, man feels an alien, a stranger. His exile is without remedy since he is deprived of the memory of a lost home or the hope of a promised land.

"This, unfortunately, is true," Wilson says, closing out the book that would mark a turning point in his life. "But we still have another hundred years."

Sociobiology drew major attention even before it was published. In a below-the-fold feature on the front page of the 28 May 1975 *New York Times,* the science reporter Boyce Rensberger wrote that word of the book's late-June publication "has spread far among biologists" and had "already stimulated considerable excitement." When Rensberger interviewed Wilson for the story, he asked him specifically if he believed that all human behavior was genetically driven. In response, he wrote, Wilson "emphasized that he was not suggesting that all human social behavior is tightly controlled by the genes." Rensberger continues, paraphrasing Wilson: "Man's intelligence . . . has enabled him to develop complex social relationships that lead to many kinds and degrees of moral commitments beyond those of other animals."

For this story, Rensberger also interviewed Robert Trivers, Wilson's colleague at Harvard and the biologist who first postulated the phenomenon of reciprocal altruism. Trivers told the journalist, "Ten years from now, the training of sociologists will have to include genetics and evolutionary theory. I'm sure sociobiology will come to have enormous impact on sociology and psychology." A drawing from the book of a pod of dolphins breaching together, captioned, "Helping wounded comrade with harpoon protruding from tail to swim to the surface and breathe," accompanied a photograph of a youthful, dark-haired Wilson looking shadowed and serious. He was weeks shy of his forty-sixth birthday.

Several politically active intellectuals in Cambridge, Massachusetts, reading the *New York Times* story, took more notice of Rensberger's assessment of the import of sociobiology than of Wilson's demurrer, and they were greatly distressed. Sociobiology, Rensberger had written, "carries with it the revolutionary implication that much of man's behavior toward his fellows, ranging from aggressive impulses to humanitarian inspirations, may be as much a product of evolution as is the structure of the hand or the size of the brain." The Cambridge observers, dedicated Marxists among them, quickly formed a collective, which they called the Sociobiology Study Group, and determined to counter what they considered Wilson's damning political alignment with eugenics.

Two prominent members of the group, the population geneticist Richard Lewontin and the paleontologist Stephen Jay Gould, were colleagues of Wilson at Harvard; Lewontin was even the chairman of Wilson's department. In response to Rensberger's story, Lewontin called the journalist and proposed that he write in turn about the evils of sociobiology. Rensberger dismissed the proposal on journalistic grounds: "There's no controversy yet," he told Lewontin. Lewontin decided to stir some up.

Sociobiology was published in late June 1975. It garnered generally favorable reviews. The science journalist John Pfeiffer, writing in *The New York Times Book Review* of 27 July, thought it "an outstanding survey addressed primarily to students and scientists but including much that will inform and intrigue serious lay readers prepared to thumb past the more formidable technical sections." A physician, Marjorie C. Meehan, in the *Journal of the American Medical Association* on 1 September, called it "well-written, coherent, a delight to read." She was "astonish[ed] that this enormous and excellent book is the work of one author." The Princeton evolutionary biologist John Tyler Bonner, in a review written that summer for *Scientific American,* first praised Wilson's *The Insect Societies* before assessing his new book as "equally impressive in its simplicity and clarity, in its enormous intelligence, even in its size. . . . It begins a new field that undoubtedly will hold a central position in biology, and perhaps in sociology too, for many years to come. It is quite an extraordinary beginning."

In the meantime, the Sociobiology Study Group was adding to its membership. By the end of the summer, it counted no fewer than sixteen Boston-area academics, professionals, and students, including, in addition to Lewontin and Gould, the Harvard Medical School microbiologist Jon Beckwith; the MIT psychologist Stephen Chorover; the Harvard biologist Ruth Hubbard; the Boston University anthropologist Anthony Leeds; and a Brandeis University premed student, Elizabeth Allen; as well as a public-school teacher, a research assistant, a resident fellow, a graduate student, a physician, and a psychiatrist. Despite its size and the prestige within academic circles of some of its members, the group had not succeeded in stirring public interest in calling out Wilson and his ambitious book. A "screen of approval" protected it, Hubbard would tell a historian.

Then the group saw its opportunity: a lengthy review of *Sociobiology* that appeared on 7 August in the twelve-year-old *New York Review of Books,* a tabloid-format periodical founded during the long New York newspaper strike of 1962–63. In the years since its founding, the *NYRB* had become a leading source of intellectual and political debate in America; Robert Silvers, its editor, would say it was committed to "a political analysis of the nature of power in America—who had it, who was affected."

The *NYRB* review, by the senior English biologist Conrad Waddington, who coined the term "epigenetics" as a plural noun ("the science concerned with the causal analysis of development") and who had his own troubles with the biological establishment, was generally favorable. Waddington harrumphs at Wilson's invocation of Camus at the beginning and end of the book, ridiculing the Frenchman for "talking Gallic rhetoric through his hat." Once he settles down to consider Wilson's work, he finds it informed with "an extraordinarily ambitious aim" that "Professor Wilson has been astonishingly successful in achieving." This aim Waddington takes to be the extension of the evolutionary Modern Synthesis to social behavior. And of Wilson's astonishing success: "This book will undoubtedly be for many years to come a major source of information about all aspects of our knowledge of the social behavior in animals, from the most primitive types such as corals, through insects, fishes, birds, to the many varieties of mammals and primitive man."

Having praised the book so lavishly, Waddington then proceeds to quarrel with its dominant theme of reciprocal altruism as the fundamental basis for social development. "That brings one," he writes, "to what is, to my mind, the weakest feature in the whole grand structure which Wilson has built up. Is it not surprising that in a book of 700 large pages about social behavior there is no explicit mention whatever of mentality?" Waddington checks the index, finding "no mention of mind, mentality, purpose, goal, aim, or any word of similar connotation." Because of this omission, he says, he feels "bound to come to the conclusion that the sociobiologists are just 'running scared' of ferocious philosophers. A few years ago it may have been tactically wise for quiet behavioral scientists to practice their own distraction procedures against the threat of predatory positivists, but I doubt if there is any longer any need for such super-caution." Waddington cannot have known how wrong he was.

A letter of protest from the Sociobiology Study Group arrived at the offices of the *NYRB* sometime in October. Signed by all sixteen members of the SSG, their names listed in alphabetical order and without group affiliation, it ran to 1,835 words under the headline "Against Sociobiology." The *NYRB*'s usual policy was to ask the reviewer to respond to such correspondence, but unfortunately Conrad Waddington had died in the interim, a second heart attack, this one fatal, outside his house in Edinburgh, on 26 September 1975. The SSG letter would stand alone, seemingly carrying the *NYRB*'s endorsement, demanding an answer.

The 13 November *NYRB* issue with the SSG's letter arrived at newsstands on 3 November, a Monday. Wilson first heard of it from his editor at the Harvard University Press, who called him "to say that word about it was spreading fast and might prove a sensation."

Sensation it was, and for Wilson a painful one. The letter labeled sociobiology "deterministic" and lumped it in with the "biological determinism" of the old eugenicists who claimed "that a host of examples of 'deviant' behavior—criminality, alcoholism, etc.—are genetically based." Though *Sociobiology* explicitly rejects the very concept of race as applied to human beings, the letter nevertheless associated Wilson's new science with a recent scandal over supposedly genetically based "racial differences in intelligence" that had snared the Stanford educational psychologist Arthur Jensen and his Nobel-laureate physicist colleague William Shockley, the co-inventor of the transistor. Seeking a motive for Wilson's perfidy, the SSG accused him in so many words of an unacknowledged political bias, of toadying to the rich and powerful:

> The reason for the survival of these recurrent determinist theories is that they consistently tend to provide a genetic justification of the status quo and of existing privileges for certain groups according to class, race or sex. Historically, powerful countries or ruling groups within them have drawn support for the maintenance or extension of their power from these products of the scientific community.

Even more brutally, the letter links Wilson's "particular theory about human nature, which has no scientific support," with those that "provided an important basis for the enactment of sterilization laws and restrictive immigration laws by the United States between 1910 and 1930 and also for the eugenics policies which led to the establishment of gas chambers in Nazi Germany." Then, after a summary analysis of Wilson's arguments, it concludes: "Wilson joins the long parade of biological determinists whose work has served to buttress the institutions of their society by exonerating them from responsibility for social problems. From what we have seen of the social and political impact of such theories in the past, we feel strongly that we should speak out against them."

Lewontin called Rensberger back when the SSG letter appeared in the *NYRB*. "*Now* there's controversy," he gloated.

Wilson remembers feeling "blindsided" by this attack, as who would not? The reviews and news reports of his book had been otherwise either positive or only critical of issues of science. "I thought there would be accolades," he told an interviewer as recently as 2019, "because [*Sociobiology*] would add to the social sciences a new armamentarium of background information, comparative analysis, terminology and general conception that could illuminate previously unexamined aspects of human social behavior."

More deeply and hurtfully, Wilson was "struck by self-doubt," which he attributes to the fact that several of the letter's signatories were members of the Harvard faculty, and two of them departmental colleagues. Lewontin's office was located directly below his in the Museum of Natural History; the plotting that led to the *NYRB* letter and later, lengthier critiques went on under Wilson's feet even as he and the plotters passed one another in genial address on the stairway. "Had I taken a fatal intellectual misstep," he worried, "by crossing the line into human behavior? . . . I faced the risk, I thought, of becoming a pariah—viewed as a poor scientist and a social blunderer to boot."

It was all too much like Wilson's long conflict with Jim Watson over the very survival of evolutionary biology. He had revealed his continuing commitment to that cause in describing his goals for sociobiology, writing in the first chapter of his weighty book that he hoped the science he was proposing to explore would contribute to seeing evolutionary biology "ranked as coordinate with such disciplines as molecular biology and developmental biology." Even more vulnerably, he had exposed his large, indeed overweening, ambition to see "sociobiology and the other social sciences, as well as the humanities," subsumed under biology.

Wilson had pushed all his chips to the center of the table. Doing so was essential to his entrepreneurial view of scientific practice—taking chances, breaking through comfortable but derelict paradigms—and he believed that the accumulating evidence justified the risk. But in the excitement of opening up an encompassing new field of science, he had forgotten the lesson of his battle with Watson: of the cynicism and viciousness of academic politics. "Controversies involving sensitive political issues," generalizes a historian of this particular conflict, Ullica Segerstråle, "exhibit something of the social psychology of witch hunts."

Wilson had exposed himself further by claiming more territory than he could yet scientifically defend: he not only wanted to subsume "the humanities" under evolutionary biology; he wanted to situate religion there as well. ("Contrary to his critics' belief," Segerstråle writes, "it was not a conservative political desire to support the existing social order that was driving Wilson. It was rather his wish to make scientific materialism triumph over irrational religious dogma that made him state his case so strongly and even exaggerate the power of evolutionary biology.") That seeming grandiosity made him vulnerable to those who were jealous of his burgeoning public reputation, as well as to political zealots, even as it played on his own insecurities as an Alabamian whose background left him an outsider in the snobbish Cambridge intellectual milieu. "I was not even an intellectual in the European or New York–Cambridge sense," he would emphasize. Despite his extraordinary achievements, then and later, Wilson never felt entirely at home in Cambridge or at Harvard, and later in life he would find his loyalties returning to the University of Alabama even as he sometimes wished he had taken up the invitation from Stanford that had moved Harvard in 1958 finally to offer him a tenure-track appointment.

Gould had at least tried to warn Wilson, cautioning him after a chance encounter in the spring of 1975, before Rensberger's front-page *New York Times* story, that there might be "political upheaval" when *Sociobiology* was published. Wilson had invited him to drop by his office to discuss the problem, but Gould, soon to join the SSG, had never followed up. Lewontin, a man of notable arrogance, assessed the conflict more contemptuously: "Wilson," he told someone, "like most scientists, expects to be able to put out a lot of bullshit about society and not get taken up on it."

Despite his bluster, Lewontin avoided confronting his older colleague face-to-face. Wilson, along with Ernst Mayr, had strongly supported Lewontin's appointment at Harvard in 1972 when there was resistance from senior faculty members who thought the population geneticist was too outspokenly Marxist to fit in. "I was had," Wilson would later say laughingly about his queries of Lewontin's colleagues at the University of Chicago on just this issue; they reassured him, he told Segerstråle, that Lewontin would be able to keep politics and science separate at Harvard. Wilson had previously supported Gould's appointment as well, in both cases to buttress his department's strength in the molecular wars. "In 1975," Wilson sums up his vulnerability, "I was a political naïf."

His anxiety soon gave way to anger. "I rethought my own evidence and logic," Wilson recalls. "What I had said was defensible as science. The attack on it was political, not evidential. The Sociobiology Study Group had no interest in the subject beyond discrediting it. They appeared to understand very little of its real substance." In August, he had written a popular preview of his book for *The New York Times Magazine*, which had published it on 12 October, in time for the SSG to have seen it before drafting their *NYRB* letter; in it he had specifically disavowed any "genetic bias." When such bias is demonstrated, he wrote there, "it cannot be used to justify a continuing practice in present and future societies." A tendency to warfare might be in our genes, he offered as an example, "but it could lead to global suicide now. . . . Our primitive old genes will have to carry the load of much more cultural change in the future. . . . Human nature can adapt to more encompassing forms of altruism and social justice." In *Sociobiology* he had been even more explicit: "What has evolved," he had written there, "is the overwhelming capacity for culture." Now, angry at the November attack, he sat down and composed a letter of response, which the *NYRB* published in its 11 December 1975 issue under the title "For Sociobiology."

Wilson said he wrote to protest "the false statements and accusations" of the SSG letter, which he called "a partisan attack." He said he resented the "ugly, irresponsible, and totally false accusation" that his book was an attempt to reinvigorate the eugenics theories and policies "which led to the establishment of gas chambers in Nazi Germany." He quoted from his *New York Times Magazine* essay to demonstrate his position that human behavior was derived 90 percent culturally and only 10 percent genetically. He protested the group's "self-righteous vigilantism," which, he said, "not only produces falsehood but also unjustly hurts individuals and through that kind of intimidation diminishes the spirit of free inquiry and discussion crucial to the health of the intellectual community."

Rensberger, in the meantime, had published the follow-up story for which Lewontin had campaigned. Much to the Harvard geneticist's frustration, the story identified him as the ringleader of the SSG attack. The reason it did so was that Lewontin, in response to Rensberger's request for information from the SSG following the *NYRB* letter and Lewontin's gloating call, had drafted the lengthy position paper Rensberger relied on for background in his follow-up story. Once again, Lewontin had been caught out attacking Wilson behind his back.

Ironically, as several scholars have noted, the conflict between Wilson and the Sociobiology Study Group was the opposite of what it seemed. Buried beneath the classic rubble of scholarly attack in the service of career ambition lay a more fundamental disagreement between traditional liberalism and the emerging radicalism of the Vietnam and post-Vietnam era. The SSG and its larger affiliation, Science for the People, had emerged from the 1960s New Left as activist groups favoring multiculturalism, the beginning of the movement in support of what the cultural historian Neil Jumonville calls "significant multicultural differences to be preserved and honored between races and ethnicities"—that is, identity politics. Wilson's Southern liberal perspective was in fact closer to that of Martin Luther King, Jr., favoring integration within a harmonious community; as King described it in his 1963 "I Have a Dream" speech, "a dream that one day on the red hills of Georgia the sons of former slaves and the sons of former slave owners will be able to sit down together at the table of brotherhood." Wilson, Jumonville argues, "was reviled by the SSG and others in the multicultural camp precisely because he denied that there were significant multicultural differences to be preserved and honored between races and ethnicities."

(In 2015, in the scientific journal *Nature Genetics,* a study appeared that settled the argument about how much of human behavior is nature and how much nurture. The study was titled "Meta-analysis of the Heritability of Human Traits Based on Fifty Years of Twin Studies." Twin studies typically compare heritability between identical twins, which are genetically identical, and sibling twins, which are no more alike than brothers or sisters. The *Nature Genetics* study looked at 7,804 traits—"virtually all human traits investigated in the past 50 years"—reported in 2,748 publications that investigated 14,558,903 twin pairs. The conclusion of this massive study: "Across all traits the reported heritability is 49%." Which means that about half of human behavior is inherited, half learned—considerably more inherited than Wilson's earlier estimate of 10 percent.)

Wilson had worked his way through anxiety to anger, the anger driving his astringent response to the SSG's *NYRB* attack. Now he found his anger subsiding. His old confidence returned, he writes, and then, characteristically, came "a fresh surge of ambition. There was an enemy in the field. An important enemy. And a new subject—which, for me, meant opportunity." Initially, that new subject was Marxism, which he set out to study. Before long, he had plunged into a broad survey of the literature of humanism, the body of knowledge he hoped sociobiology in the fullness of time would subsume. The result of that deep reading would be a new book that for Wilson was also a new kind of book: an examination of sociobiology from a broader, humanistic perspective; as he explains in its preface, "not a work of science [but] a work about science." He researched and wrote it in 1976 and published it in 1978. He called it *On Human Nature.*

10

The Deep Things

SEEKING A BROADER AUDIENCE than the science professionals for whom he had written *Sociobiology*, Ed Wilson begins *On Human Nature* by rephrasing Albert Camus's tragic vision of human alienation in "a universe suddenly divested of illusions." This time, having enlarged his perspective with guided reading through the literature of humanism, Wilson speaks in his own voice, with confident authority, as he will speak throughout the book.

Now he finds not one but two "great spiritual dilemmas." Camus's is only the first. Wilson anchors the French writer's existential dilemma in biology:

> The reflective person knows that his life is in some incomprehensible manner guided through a biological ontogeny, a more or less fixed order of life stages. He senses that with all the drive, wit, love, pride, anger, hope, and anxiety that characterize the species he will in the end be sure only of helping to perpetuate the same cycle. Poets have defined this truth as tragedy. . . .
>
> The first dilemma, in a word, is that we have no particular place to go. The species lacks any goal external to its own biological nature. It could be that in the next hundred years human-

kind will thread the needles of technology and politics, solve the energy and materials crises, avert nuclear war, and control reproduction. The world can at least hope for a stable ecosystem and a well-nourished population. But what then?

Thus the first dilemma—the end of illusion, the darkening of our species' presumptions of millennial purpose to the brutal view T. S. Eliot's louche character Sweeney declares in Eliot's 1932 play *Sweeney Agonistes:*

SWEENEY:
. . . Birth, and copulation, and death.
That's all the facts when you come to brass tacks:
Birth, and copulation, and death.

Then Wilson, who is consistently impatient with philosophers who anchor their ideas in intuition alone, moves beyond Camus to identify humanity's second spiritual dilemma. Since it's more complicated to explain, he only summarizes it at first, promising to explore it in detail later. At the same time, he quietly plants a land mine for the Sociobiology Study Group, incorporating an implicit challenge to the group's claims in his initial précis:

> Innate censors and motivators exist in the brain that deeply and unconsciously affect our ethical premises; from these roots, morality evolved as instinct. If that perception is correct, science may soon be in a position to investigate the very origin and meaning of human values, from which all ethical pronouncements and much of political practice flow.

In Wilson's view, we aren't blank slates. Human beings are born with a network of unconscious biological constraints, shaped by evolution, that limit the range of our basic values. If so, and assuming science advances in understanding how the brain works, the time may come ("soon," Wilson predicts) when the biological origins of our values become evident. That success would generate our second spiritual dilemma: having to decide "which of the [biological] censors and motivators should be obeyed and which ones might better be curtailed or sublimated." Even more fundamentally: "At some time in the future we will have to decide how human we wish to remain."

Deciding what to keep and what to suppress of our biological endowment would be eugenics, however enlightened: here is Wilson's implicit challenge to his SSG tormentors, who accuse him of peddling supposedly unscientific claims to support the status quo. Continue scientific research, he's arguing, and the time will arrive when sorting out our humanity becomes inevitable (just as the time came in physics, with the discovery of nuclear fission, when historical circumstances made it impossible not to release nuclear energy for good and for ill, with world-transforming consequences that are still unfolding).

Science works by gift exchange: publishing a discovery makes it available to other scientists to use as an instrument or as a clue to make further discoveries, to be published in their turn. Suppress the research and you kill science; kill science and you lose its evident benefits, both material and moral, and accept continuing harm. Nor are those goods and ills evident at the outset, when it might be possible to choose among them. As Wilson's hero Robert Oppenheimer put it in a 1962 lecture, "The deep things in science are not found because they are useful; they are found because it was possible to find them."

To drill his point through, Wilson sharpens it:

> The only way forward is to study human nature as part of the natural sciences, in an attempt to integrate the natural sciences with the social sciences and humanities. I can conceive of no ideological or formalistic shortcut. Neurobiology cannot be learned at the feet of a guru. The consequences of genetic history cannot be chosen by legislatures. Above all . . . ethical philosophy must not be left in the hands of the merely wise. Although human prog-

> ress can be achieved by intuition and force of will, only hard-won empirical knowledge of our biological nature will allow us to make optimum choices among the competing criteria of progress.

Much of *On Human Nature* is devoted to marshaling evidence of hereditary influence on human behavior. Wilson the synthesizer draws on his broad reading of the scientific literature to make his case, incidentally demonstrating how passionately ill-informed his challengers were. An American anthropologist's 1945 listing of characteristics common to "every culture known to history and ethnography" includes "age-grading, athletic sports, bodily adornment, calendar, cleanliness training, community organization, cooking, cooperative labor, cosmology, courtship, dancing, decorative art, divination, division of labor, dream interpretation, education, eschatology, ethics," and so on through the alphabet for half a page, concluding with "sexual restrictions, soul concepts, status differentiation, surgery, tool making, trade, visiting, weaving, and weather control."

For contrast, Wilson then compiles a similar list of characteristics one would find in a society of intelligent ants, including "age-grading, antennal rites, body licking, calendar, cannibalism, caste determination, caste laws, colony-foundation rules, colony organization, cleanliness training, communal nurseries, cooperative labor, cosmology, courtship, division of labor, drone control," concluding with "soldier castes, sisterhoods, status differentiation, sterile workers, surgery, symbiont care, tool making, trade, visiting, weather control, and still other activities so alien as to make mere description by our language difficult."

Not only do these lists point to genetic differences unexplainable by social advancements or intelligence alone; they also demonstrate, as Wilson writes, that human nature "is just one hodgepodge out of many conceivable"—another indication of the strong influence of random Mendelian mutation—of DNA copying mistakes—on its development.

Wilson isn't arguing for a simple correspondence between a single gene and a social trait—"there will be no mutations," he writes, perhaps puckishly, "for a particular sexual practice or mode of dress." The complexity of social behavior points instead to multiple genes and subtle effects: "The behavioral genes more probably influence the ranges of the form and intensity of emotional responses, the thresholds of arousals, the readiness to learn certain stimuli as opposed to others, and the pattern of sensitivity to additional environmental factors that point cultural evolution in one direction as opposed to another."

As he works his way in *On Human Nature* through heredity, development, emergence, aggression, sex, and altruism, Wilson arrives finally at religion. "The predisposition to religious belief," he opens that penultimate chapter, "is the most complex and powerful force in the human mind and in all probability an ineradicable part of human nature." Whether that is so remains to be seen, but there is certainly a case to be made, and Wilson proceeds to make it, not only to investigate religion's biological roots but also to explore his own religious experience and the secular religion of his radical challengers. One staggering note at the outset: "According to the anthropologist Anthony F. C. Wallace, mankind has produced on the order of 100 thousand religions."

Such a universal production is unexplainable unless it seeks to meet a deep need. Wilson is impatient with scientific humanists who dismiss religion as superstition, who "organize campaigns to discredit Christian fundamentalism, astrology, and Immanuel Velikovsky." (Velikovsky was a Lithuanian Jewish psychiatrist and cosmic theorist who published best-selling books in the 1950s and 1960s claiming that the Biblical flood and other Biblical and ancient historical events had real causes in spectacular planetary mass ejections and near-collisions.) Their "crisply logical salvos," he mocks them, "endorsed by whole arrogances of Nobel Laureates, pass like steel-jacketed bullets through fog." Wilson can be impatient with impertinence, but the level of his ridicule here follows from his personal experience of religion, however abortive, his empathy for those who practice it—and his memories of encounters with the arrogance of his nemesis, Jim Watson. "The humanists are vastly outnumbered by true believers," he points out, adding, "Men, it appears, would rather believe than know." The assertion is one of his touchstone phrases, reappearing in several of his texts; no other claim angered the SSG members more.

Two important caveats qualify Wilson's examination of the sociobiology of religion: first, that "by traditional methods of reduction and analysis science can explain religion but cannot diminish the importance of its substance"; second, that, because religion is unique to the human species, principles of behavioral evolution drawn from animal studies are unlikely to apply to it directly.

Wilson analyzes religion in more extended and multilevel detail than is possible here. At its heart, he finds it to be humankind's most powerful support for altruism: "The votary is ready to reassert allegiance to his tribe and family, perform charities, consecrate his life, leave for the hunt, join the battle, die for God and country." It is as well a system for codifying social rules, an argument Wilson attributes to the early-twentieth-century French philosopher Henri Bergson. "The extreme plasticity of human social behavior," Wilson writes, citing Bergson, "is both a great strength and a danger. If each family worked out its own rules of behavior, the society as a whole would disintegrate into chaos. To counteract selfish behavior and the dissolving power of high intelligence and idiosyncrasy, each society must codify itself." Ironically, almost any set of rules will do, as the hundred thousand crazy-quilt variety of religions makes clear.

What works for the group also works for the individual. "The highest forms of religious practice," Wilson writes, "congeal identity"—crystallize it from a fluid state into a solid:

> In the midst of the chaotic and potentially disorienting experiences each person undergoes daily, religion classifies him, provides him with unquestioned membership in a group claiming great powers, and by this means gives him a driving purpose in life compatible with his self-interest. His strength is the strength of the group, his guide the sacred covenant.

Why Wilson assigns "the highest forms of religious practice" to this duty isn't obvious, especially since his description of the benefits of individual belief prepares the ground for his examination of the dogma of the Sociobiology Study Group and its affiliates.

If religious belief congeals identity, dogma curdles it. Wilson finds three great mythologies contending in the contemporary world: "Marxism, traditional religion, and scientific materialism." An acknowledged scientific materialist, he has some respect for traditional religion but very little for Marxism as he has encountered it:

> Marxism is sociobiology without biology. The strongest opposition to the scientific study of human nature has come from a small number of Marxist biologists and anthropologists who are committed to the view that human behavior arises from a very few unstructured drives. They believe that nothing exists in the untrained human mind that cannot be readily channeled to the purposes of the revolutionary socialist state. When faced with the evidence of greater structure, their response has been to declare human nature off limits to further scientific investigation. A few otherwise very able scholars have gone so far as to suggest that merely to talk about the subject is dangerous, at least to their concept of progress. I hope I have been able to show that this perception is profoundly wrong.

More than wrong, as Wilson concludes with a bold challenge: Marxism, "formulated as the enemy of ignorance and superstition," has gone dogmatic "and is now mortally threatened by the discoveries of human sociobiology."

On Human Nature, published in 1978, is a relatively late product of Wilson's response to the attacks he encountered when he published *Sociobiology* in 1975. It was typical of his view of how science is done (and how honorable colleagues are supposed to behave) that he would write a book that he hoped would explain his ideas more clearly, and thus resolve any disagreements.

After he responded to the Sociobiology Study Group's initial attack in *The New York Review of Books,* Wilson began to be harassed at Harvard and at several of the conferences he attended from 1976 to 1978 to discuss and defend his "new synthesis." The Harvard harassment worried him most, because he feared it would reach his home in Lexington and involve his wife, Irene, and his daughter, Cathy, who was now entering adolescence.

"I have wavered about going to several lectures," he told the journalist Nicholas Wade in February 1976. "There has been clearly prearranged hostile questioning. Perhaps a braver soul would not have been concerned, but I find it intimidating." He withdrew from a public talk scheduled in late March, Wade reported, "because of the increasing mental strain on his family." He told the novelist Michael Crichton, with whom he had dinner during this period, "At times the protest reached the level of interruption of my classes and public demonstrations. One in Harvard Square demanded my dismissal from Harvard." (This group, the International Committee Against Racism—InCAR, the activist front organization of the small, radical Progressive Labor Party in the United States—picketed and handed out leaflets urging demonstrations against Wilson and, as he told Crichton, demanding he be fired. It would not be his last encounter with them.) The novelist asked him how he handled the pressure. "It was embarrassing at times for me and my family," he said, "but intellectually not difficult. It was obviously a contest of science against political ideology, and past history has shown that if the research is sound, science eventually comes out on top."

The two conferences Wilson and others remember most vividly involved the famous anthropologist Margaret Mead in the first instance, and a pitcher of ice water in the second.

Anthropologists had not taken kindly to Wilson's book. Not only were they professionally hostile to genetic explanations of behavior, which they viewed as culturally derived; they also resented Wilson's bold claim, as the anthropologist Napoleon Chagnon (pronounced SHAG-non) writes, that sociobiology "would most likely become the overriding and comprehensive theoretical framework in the life sciences, one that would subsume other sciences—including anthropology and sociology—as subordinate components."

Unlike most anthropologists, however, Chagnon and his Penn State colleague William Irons found much of interest and much to admire in Wilson's new synthesis. Chagnon in particular was intrigued with Bill Hamilton's work on kin selection, he writes, "because it laid a new basis for understanding why kinship relationships provided the rock-bottom source of social solidarity." To explore Wilson's ideas, the two anthropologists, along with several colleagues, organized a large block of formal sessions on sociobiology for the seventy-fifth annual meeting of the American Anthropological Association, which would be convened in Washington, D.C., during the third week of November 1976. There would be three half-day sessions on sociobiology on Friday and two more on Saturday, for a total of twelve hours of presentations and discussion, a large time commitment for one subject during a four-day international conference.

The Harvard anthropologist Irven DeVore would chair the major session, on Friday morning, with speakers including Wilson, Robert Trivers, and the young primatologist Sarah Blaffer Hrdy, who had just earned her Ph.D. under DeVore, with Trivers and Wilson on her thesis committee. The psychiatrist David A. Hamburg, at that time president of the Institute of Medicine of the National Academy of Sciences, later to become president of the Carnegie Corporation, would respond to the Friday-morning presentations as discussant.

Some fifteen hundred anthropologists and scientists in related disciplines from throughout the United States and abroad, including Margaret Mead, gathered at the Washington Hilton in the coldest November in more than twenty years. Washington was astir; Jimmy Carter had just defeated Gerald Ford in a presidential election.

Although Wilson and his supporters were not scheduled to speak until Friday, a lecture highlighting the opening-night plenary session on Wednesday evening glared a major challenge. Earlier that year, the speaker, the University of Chicago anthropologist Marshall Sahlins, had published a slashing critique of sociobiology, *The Use and Abuse of Biology,* attacking Wilson's proposed new science as fundamentally deficient. As a reviewer summarizes Sahlins's argument, "He declares that culture is so independent that it cannot be studied in reference to anything but itself, and he sets out to annihilate the entire notion that natural selection could possibly have anything new to say about why humans act as they do." The reviewer was right to choose the word "annihilate"; as Sahlins himself quips in his book, "We . . . indulge our aggressions and commit mayhem by writing books and giving lectures." Since Sahlins's book had made his hostility to sociobiology public knowledge, it was no accident that he was chosen to speak to the association on the opening night of its annual meeting; whoever had scheduled him was clearly opposed at least to sociobiology's dominance of the meeting schedule.

The real fireworks burst on the following evening, in advance of the sociobiology sessions beginning Friday morning. The evening plenary session was supposed to be a business meeting. It was instead nearly a public hanging. Nap Chagnon recalls it vividly in his memoir *Noble Savages,* which is pointedly subtitled *My Life Among Two Dangerous Tribes—the Yanomamö and the Anthropologists:*

> The ballroom in which the business meeting was held was full beyond capacity. A motion had been placed on the agenda by opponents of sociobiology aimed at preventing the sessions that Irons and I had organized. . . . It was as if the last two bastions of opposition to the theory of evolution by natural selection were fundamentalist fire-and-brimstone preachers and cultural anthropologists! Heated debate and impassioned accusations of racism, fascism, and Nazism punctuated the frenzied business meeting that night.

When Chagnon had listened to as much as he could stomach, he moved to table the motion. "Nobody heard me," he remembers, "because Margaret Mead, the 'Mother Goddess' of anthropology, stood up to address the motion to prohibit our sessions." Mead was a former president of the association and had just ended a year as president of the American Association for the Advancement of Science. The room went silent:

> She began by expressing her opposition to Ed Wilson's book and his whole idea of a "science of sociobiology" that would possibly subsume anthropology as a subordinate discipline. However, she said, in spite of her opposition to Wilson's book and to sociobiology, she felt that the motion as worded was essentially a "book burning" motion and, for that reason, she thought that it was not something our association should advocate and be identified with. She then sat down, somewhat regally, and the vote on the motion was taken almost immediately. The motion was defeated, but not by a wide range.

Mead was seventy-four, a cultural anthropologist of world celebrity. Her opposition to sociobiology was evidently professional, founded on her rejection of Wilson's claim that sociobiology would incorporate her scientific field. A year later, she invited Wilson to dinner, specifically to discuss sociobiology, during a conference they both happened to be attending in Virginia. "I was nervous then," he writes, "expecting America's mother figure to scold me about the dangers of genetic determinism." But the great anthropologist had decided instead to claim priority. "I had nothing to fear. She wanted to stress that she, too, had published ideas on the biological basis of social behavior. . . . Over roast beef and red wine . . . she recommended several of her own writings that she thought I might want to read."

In 1976, Wilson was awarded a National Medal of Science, the nation's highest scientific honor. One of fifteen recipients, he received the award from President Jimmy Carter on the morning of 22 November 1977, almost exactly a year after the contentious AAA annual meeting, in a ceremony held in the Old Executive Office Building. A twelve-member committee of the National Science Foundation had chosen the awardees; Wilson's award citation credited him "for his pioneering work on the organization of insect societies and the evolution of social behavior among insects and other animals." If there were anthropologists on the NSF committee, their opinion hadn't prevailed. The award, Wilson told me, was one important factor in the fading away of public controversy about sociobiology as the seventies wound down. Another was the success of his book *On Human Nature,* published in 1978 and awarded a Pulitzer Prize in general nonfiction the following year—Wilson's first Pulitzer, but not his last.

Between those years, a less noble savagery marred the 1978 annual meeting of the American Association for the Advancement of Science, held on five days in mid-February in Washington. According to Chagnon, after the infighting at the 1976 anthropology meeting, campuses across the United States had sponsored debates about sociobiology, "the general theme [of which] was something like 'What's Dangerous About Sociobiology?'" By that time, Chagnon writes, "the very word *sociobiology* had become a lightning rod in the social sciences on college campuses." Chagnon had participated in some of the debates, as had Bob Trivers. Always outspoken, the irascible Trivers had told Chagnon at one point, "I've finally figured out what they mean by a 'balanced' debate. For every clear demonstration of how effective a sociobiological explanation is of some phenomenon, it must be 'balanced' by a completely nonsensical appeal to bullshit, emotions and political correctness."

In response to these acrimonious campus debates, Chagnon recalls, the AAAS decided to sponsor a national debate on sociobiology at its 1978 annual meeting, "hoping to put an end to the squabbling and to the unreasonable and false accusations against those who wished to study human behavior with new theoretical insights into natural selection." A biologist and an anthropologist served as conveners, inviting leading proponents and opponents to participate. Margaret Mead was scheduled to serve as moderator—that was the reason she had invited Wilson to dinner not long before the event—but a diagnosis of pancreatic cancer forced her to withdraw. She died of that implacable disease the following November.

A full two-day program, "Beyond Nature-Nurture," culminated in a symposium on Wednesday afternoon, 15 February 1978, where Wilson was to appear as the final speaker. Chagnon and Bill Irons

Ants sustained in lab colonies enabled EOW to discover the glandular secretions they use to communicate and to "talk" to them. He was the first in the world to do so.

EOW and Robert MacArthur had studied the 1883 eruption of Krakatau to learn how islands repopulate after disaster. MacArthur's early death at forty-two in 1972 ended a legendary career.

W. D. Hamilton (l) and Robert Trivers (r) contributed their pioneering theories of the evolution of social behavior to EOW's comprehensive synthesis of sociobiology.

EOW's Harvard colleague Richard Lewontin attacked sociobiology from a Marxist perspective. The battle was brutal.

The world-renowned anthropologist Margaret Mead disliked sociobiology but defended EOW's right to speak when her colleagues were voting to condemn him.

Controversial anthropologist Napoleon Chagnon, who had endured similar attacks for his work in the Amazon, saw value in sociobiology and rushed to EOW's defense.

EOW returned to the field in the 1980s as an antidote to the bitter political debates he had endured. The increasing pace of species extinctions called him to activism.

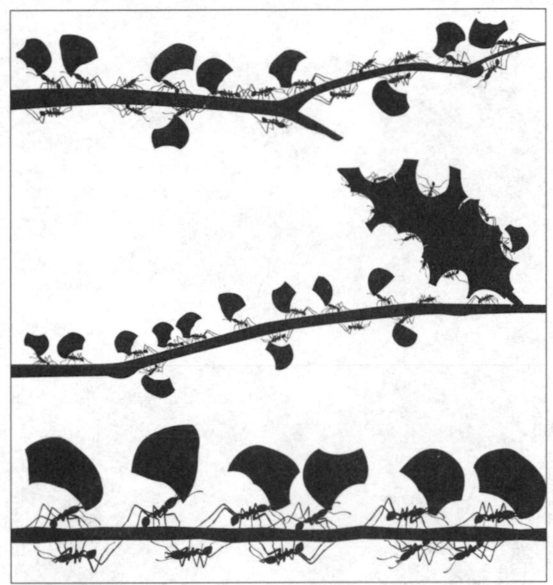

Social successes despite their small brains, ants range from fierce predators to careful farmers. Leafcutter ants use organic matter to grow fungus, which they in turn consume.

EOW and Bert Hölldobler won a Pulitzer Prize in 1990 for their eloquent compendium *The Ants*. Wilson assistant Kathy Horton loyally supported EOW's long career.

As EOW's insights into the natural world enlarged across the years, he never forgot "the animals that can be picked up between thumb and forefinger and brought close for inspection."

Philanthropist Greg Carr encouraged EOW to support his work rehabilitating Mozambique's magnificent Gorongosa National Park, ravaged in a long civil war.

Two towering pioneers, EOW and James Watson eventually put aside their differences and became good friends.

One of the great biologists of all time, EOW continues his work of advocacy for species preservation into his tenth decade.

had presented earlier; Chagnon had noticed "that the first two rows nearest the podium were filled with mostly sullen-looking young men who showed no signs of interest in the papers." All the presenters for and against, Stephen Gould among them, were seated in a line in chairs onstage divided by a central podium. As the meeting proceeded, detractors in the audience challenged the speakers, including both Bill Hamilton and the British ethologist Richard Dawkins, whose popular book *The Selfish Gene* had been published less than two years earlier. Hamilton responded solemnly and somewhat shyly. Dawkins climbed up on his chair, Chagnon writes, "and delivered from that commanding position well-considered and highly informed responses to the criticisms of his work."

Then it was Wilson's turn. He had come to the meeting at a physical disadvantage: on crutches, with a cast on his right ankle, from a fall he had taken on black ice while jogging in Cambridge only two weeks before. He stayed seated as he began his talk; he could not easily have stood or moved.

The "sullen-looking young men" and women in the first rows of the audience immediately rushed the stage, shouting slogans and insults, including one of their favorites: "Racist Wilson, you can't hide, we charge you with genocide!" Wilson had been cautioned before the session began that the same International Committee Against Racism that had harassed him in Harvard Square was planning a demonstration. Two InCAR members had passed out protest leaflets, refusing one to Wilson. The group was known for violent actions, and here they were.

About eight InCAR men and women, Wilson estimates, lined up behind the row of presenters. "Several held up anti-sociobiology placards, on at least one of which was painted a swastika." Their leader strode to the lectern to take the microphone. Chagnon remembers the moderator, Alexander Alland, Jr., a Columbia University anthropologist, "shouting 'Please stop! Please stop! I'm one of you people. . . . I'm also a Marxist! This is unacceptable!'" Chagnon, who was trying to push forward through the crowd of fleeing audience members to go to Wilson's aid, adds, "Maybe Marxists had some kind of secret code I didn't know about, but it didn't work: the 'Marxists' continued to attack Wilson. . . . It was the most hateful, frightening, disgusting behavior I've ever witnessed at an academic assembly."

Following AAAS procedure, Alland turned over the microphone to the InCAR leader, telling him he had two minutes before hotel security would be called. The man began haranguing the audience. Then, the most humiliating moment of all for Wilson, a young woman behind him hauled up a pitcher of ice water and drenched it over his head as the demonstrators chanted, "Wilson, you're all wet!"

The thoroughly soaked sociobiologist dried himself off as best he could with his handkerchief; someone passed him a paper towel. "In a little over two minutes," he recalls with evident anger, "they left the stage and took their seats. No one asked them to leave the premises, no police were called, and no action was taken against them later. After the symposium several stayed behind to chat with members of the audience."

A report in the AAAS journal *Science* blandly recites the rest of the story: "The moderator then made an apology to Wilson and his copanelists which prompted a standing ovation from the audience." (Wilson: "Of course they [applauded], I thought. What else could they do? They might be next.") The *Science* report concludes: "Whereupon Wilson, who had to take all of this sitting down because his ankle was in a cast, proceeded with his paper on 'Trends in Sociobiological Research.'"

For Wilson, one good result at least came out of these ugly episodes: Jim Watson invited him to keynote a conference at Cold Spring Harbor on neurobiology and behavior. "I went," Wilson recalls, "and I gave the lecture, and that's when Watson and I became good friends. He introduced me in a way that showed the generosity of his soul. He said, 'Ed Wilson is the kind of biologist who makes the discoveries we work on.'" Bruce Stillman, Watson's successor as director, said Watson told him, "Any enemy of Ed Wilson's is an enemy of mine." On two episodes of *Charlie Rose,* the talk show on PBS television, the two old enemies sparred in a friendly way but clearly liked and enjoyed each other.

Wilson wrote two more books about sociobiology, both with a young Canadian theoretical biologist, Charles J. Lumsden: *Genes, Mind, and Culture,* published in 1981, and *Promethean Fire,* published in 1983. Their common theme was the coevolution of genes and culture. In these works, they coined the term "culturgen" to refer to ideas that are culturally transmitted, a variant of Richard Dawkins's earlier and more popular term "meme." Debate continued, muted to the book-review pages of scientific journals and, inevitably, *The New York Review of Books.*

"I got tired of the sociobiology controversy," Wilson told me. "I think I won it hands down, because there were not that many outspoken critics like Lewontin and Gould, and they just grew silent, most of them. Maybe they're still murmuring in classrooms, but, no, I was happy to get back to something I felt more comfortable with. Arguing with scientists or others who are strongly against something you've done is not a pleasant occupation."

That something Wilson got back to in the years after 1980 was ecology, and particularly the disturbing and intensifying decline in biodiversity and even outright extinction that followed from increasing human exploitation of the natural world.

11

Crossing the Line

ED WILSON'S BOOK *Biophilia,* published in 1984, almost sings with the joy of returning to the natural world from the mean precincts of academic brawling. "Biophilia," love of the natural world, Wilson defines as "the rich, natural pleasure that comes from being surrounded by living organisms." He claims it among "our deepest needs," a genetic endowment, and whether it is or not, it certainly felt that way to him.

He had not set aside his scientific communications in the years of fighting for sociobiology: between 1975 and 1984, he published ninety-eight papers, essays, books, introductions, and book reviews, including his books on sociobiology (*Sociobiology; On Human Nature; Genes, Mind, and Culture; Promethean Fire*), two edited books, *Animal Behavior* and *The Insects,* and one study, *Caste and Ecology in the Social Insects,* co-written with the University of California mathematical biologist George Oster. He carried on his Harvard teaching as well, an academic star, adding a lecture course on sociobiology to his load. The self-described workaholic continued to pour forth work.

In 1980, he found a new and transformative cause: species extinction from the accelerating destruction of wildlife habitat, especially the great tropical rain forests of Central and South America. It recalled him to his visit to Cuba as a Junior Fellow in 1953, before he ventured out to the South Pacific, when he had seen the ravages of a country almost entirely given over to the cultivation of sugar cane, with only isolated shreds of forest left on hills and mountainsides too steep to cultivate. Then the destruction had seemed regional and local, not yet advanced enough globally to raise urgent concern. That was when the natural world still felt immortal to him.

In the 1960s, he had explored island biodiversity, even deliberately depopulating mangrove islands in the Florida Keys with Dan Simberloff to see how life returned. He and Robert MacArthur had researched and written *The Theory of Island Biogeography*, published in 1967, observing there that islands might be other than elevations of land surrounded by water. A forest remnant surrounded by pasture, farmland, or human settlement was an island—a "habitat island," he and MacArthur called it. They had observed as well that "a reduction of habitat is inexorably followed by a loss of animal and plant species. Very roughly, we learned, a 90 percent reduction of forest cover . . . eventually halves the number of species living there." But then he had grasped the opportunity to synthesize an entire field of study, years of independent discoveries by dozens of biologists published in scattered journals about social insects, and *The Insect Societies* diverted him, and then, through the 1970s, sociobiology.

The battering he took over sociobiology made him wary. In *Naturalist* he describes a recurring nightmare from this period, in which he fails to collect an island's riches of flora and fauna before he's forced to leave it, and from which he awakens "knotted with anxiety and regret." He wondered, as the 1970s passed, if and when scientists should become activists: "Speak too forcefully, I thought, and other scientists regard you as an ideologue; speak too softly, and you duck a moral responsibility." There were large organizations devoted to the cause of habitat destruction. Maybe the next generation of young biologists would embrace it.

A scientific report alerted him to the accelerating crisis. The U.S. National Research Council had commissioned it in 1978 from a British ecologist trained at Berkeley, Norman Myers, seeking to learn "which biological questions can be answered only in the tropics, or can be answered better or more efficiently there than elsewhere." Myers pulled together whatever documentary evidence of tropical rain-forest conversion he could find, focusing on forest farming, timber trade (solid wood and paper pulp, commercial logging, wood chips), cattle raising, and firewood cutting.

The world's rain forests, the NRC consultant found, occupied at that time about 7 percent of the world's land area—about the same as the land area of the contiguous forty-eight United States—but, in Wilson's words, "they teem with the greatest variety of plants and animals of all the world's ecosystems." And roughly .25 percent of their species were "extinguished or doomed to early extinction each year" by the annual cutting and burning worldwide of an area of forest equal to about half the area of Florida.

Myers's grim conclusion: the world was losing not one species per year, as most biologists had previously assumed, but one species or more per *day*:

> It is not unrealistic to suppose that, within perhaps the next two decades, many thousands of species could disappear in [tropical rain forest]. . . . Elimination of a substantial proportion of the planetary spectrum of species will mean a gross reduction of life's diversity on earth; and it will entail a permanent shift in the course of evolution, and an irreversible loss of economic opportunity as well.

The chair of the NRC committee was an old friend of Wilson's, Peter Raven, a botanist and former Stanford professor and since 1971 the director of the Missouri Botanical Garden in St. Louis, the second-largest such institution in North America. "A distinguished scientist," Wilson writes of him, ". . . increasingly a public figure, Peter was determined and fearless. He had no qualms about activism." During the half-decade of the later 1970s when Wilson was preoccupied with defending sociobiology, his fearless friend, he says, "was writing, lecturing, and debating those still skeptical about the evidence of mass extinction. . . . More than anyone else Raven made it clear that scientists in universities . . . must get involved; the conservation professionals could not be expected to carry the burden alone."

These accumulating challenges worked on Wilson, moving him in the direction of activism. His first public statement was brief, but it marked his debut as an environmental activist: for its first issue in 1980, *Harvard Magazine*, an independent bimonthly affiliated with its namesake university, invited seven Harvard professors to describe what they believed to be the most important problem facing the world in the new decade. "Four [of the seven]," Wilson writes, "cited poverty arising from, variously, overpopulation, the influx of rural masses into cities, and capitalism. Another, focusing on the United States, cited the welfare state and excessive governmental control. The sixth chose the global nuclear threat." None mentioned the environment. Wilson stepped up, opening his brief, column-long statement with a prediction:

> Permit me to rephrase the question as follows: What event likely to occur in the 1980s will our descendants most regret, even those living a thousand years from now? My opinion is not conventional, although I wish it were. The worst thing that can happen—*will* happen—is not energy depletion, economic collapse, limited nuclear war, or conquest by a totalitarian government. As terrible as those catastrophes would be for us, they can be repaired within a few generations. The one process ongoing in the 1980s that will take millions of years to correct is the loss of genetic and species diversity by the destruction of natural habitats. This is the folly our descendants are least likely to forgive us.

Not long after, Wilson says, he called up Peter Raven and volunteered: "One day on impulse I crossed the line. I picked up the telephone and said, 'Peter, I want you to know that I'm joining you in this effort. I'm going to do everything in my power to help.' "

Wilson immediately enlisted with a coalition of senior biologists active in working on rain-forest preservation that he dubbed the "rain forest mafia." The group included Raven, Myers, the geographer and ornithologist Jared Diamond, the Stanford biologist Paul Ehrlich, Wilson's Cornell entomologist colleague Thomas Eisner, the University of Pennsylvania ecologist Daniel Janzen, and the director of conservation for the World Wildlife Fund, Thomas Lovejoy III. Through Lovejoy, Wilson was elected to the board of directors of the WWF-US and became its external science adviser.

Wilson and several of his fellow rain-forest mafiosi, along with other concerned scientists, announced their commitment to tropical forest conservation with personal financial contributions in a letter to the U.S. journal *Science* in September 1981. They wrote of "the silent crisis of our time: species extinction," which they argued was far more frequent now "than in any recent period in geological history and is accelerating." They feared as many as a million species might be eliminated in the next quarter-century. Among those signing the letter were the astrophysicist Carl Sagan, Peter Raven, Tom Eisner, Ernst Mayr, Anne and Paul Ehrlich, and of course Wilson. "Species diversity," the letter proclaimed, "is a great treasure house where riches have not yet been closely examined, much less used by humanity. Among the still largely unknown millions of species are vast potential sources of new foods and pharmaceutical and other natural products, agents of nitrogen fixation and soil formation, defenses against insect pests, and, not least, objects of beauty, enchantment, and wonder."

Each signer pledged to contribute a thousand dollars (equivalent to three thousand today) to a fund being raised to buy a fifteen-hundred-acre tract adjacent to a two-thousand-acre Costa Rica lowland rain-forest preserve recently chosen as one of four primary sites in the world for detailed studies of tropical ecosystems. To make the case for preservation, the rain-forest mafia needed scientific knowledge of the rain-forest environment as well as of its plants and animals.

During the next several years, Wilson lectured and wrote extensively about the threat of species extinctions. At the same time, he continued his research on social insects. In that work, the two commitments came together, and in the spring and summer of 1983, he found himself doing fieldwork in an area of Brazilian rain forest where fellow scientists were deliberately clearing and burning wide lanes of rain-forest trees. Tom Lovejoy of the World Wildlife Fund had invited Wilson to visit the field site of the Minimum Critical Size of Ecosystems Project. The project was a joint effort between the WWF and Brazil's National Institute for Research on Amazonia, located on three large cattle ranches at the edge of rain forest about sixty miles north of Manaus, the capital of the state of Amazonas in western Brazil.

The Critical Size Project was looking for answers to a question that had been widely debated among conservationists in the 1970s that went by the acronym SLOSS—Single Large or Several Small—referring to how nature reserves should be structured. No one knew which arrangement would better preserve the natural diversity of a biome. Lovejoy's heroic project in Brazil intended to study the question by setting up rain-forest reserves of various sizes, from one hectare (an area about the size of two American football fields) to one hundred hectares (about ten city blocks or about a third of a square mile).

Wilson was especially welcome because his work with Robert MacArthur on island biogeography had offered Lovejoy a theoretical structure on which to ground his field studies: much of what Wilson and MacArthur had generalized from the effects of island relationships on species populations applied directly to the rain-forest islands Lovejoy and his team were creating. For Wilson, the chance to collect again promised to be restorative, especially since it allowed him to deepen his understanding of an urgent problem while carrying on his own scientific work as well.

On one August night, he took a chair and sat out in a clearing away from the camp's noise and smells. It was so dark at the edge of the rain forest that he couldn't see his hand in front of his face. He was tired and bored, he writes, working his mind through "the labyrinths of field biology and ambition," ready to be distracted. He clicked on his headlamp. "I swept the ground with the beam . . . and found—diamonds!":

> At regular intervals . . . intense pinpoints of white light winked on and off with each turning of the lamp. They were reflections from the eyes of wolf spiders, members of the family Lycosidae, on the prowl for insect prey. When spotlighted the spiders froze, allowing me to approach them on hands and knees and study them almost at their own level. I could distinguish a wide variety of species by size, color, and hairiness. It struck me how little is known about these creatures of the rain forest, and how deeply satisfying it would be to spend months, years, the rest of my life in these places until I knew all the species by name and every detail of their lives.

Elsewhere, Wilson writes of taking a Magellanic voyage—Magellan, who first circled the Earth—around a forest tree, collecting, studying, and paying homage to the life that lived on and under its bark. His nostalgia for wilderness surely connects back to his Huck Finn childhood. But any clear mind, once familiar with the surroundings, must respond to the fractality of the natural world, the realization that it is equally complex at every level, from atoms to galaxies, from viruses to rain forests to the human brain. Ed Wilson studied the wolf spiders, and the wolf spiders studied Ed Wilson: "Even these species turning about now to watch me from the bare yellow clay could give meaning to the lifetimes of many naturalists." And the rain forest that began at his feet cathedraled northward five hundred kilometers to Venezuela.

For the people who lived on and by the land, the rain forest was either a resource or an impediment. The three ranches enclosing the Critical Size Project were failing. The forest soil, that wet-desert yellow clay, was low in nutrients; almost all the forest's fertility had been drawn up into its trees, from roots to towering canopy. Clearing the trees with burning brought a flush of fertile ash to the old soil that supported at best a few years of cattle pasture or poor cropland. The deeply damaged rain forest took centuries to recover. In *Biodiversity*, a book Wilson edited for the National Academy of Sciences, he cites an example from ancient Cambodia: "The forest at Angkor . . . dates back to the abandonment of the Khmer capital in 1431, yet is still structurally different from a climax forest today, 556 years later."

So the sixty plots of rain forest that Lovejoy's teams had shaped and were studying would help determine how large Brazil's reserves should be to sustain something like the diversity of the original forest, if that was possible. If not, the reserve configuration should at least slow and limit extinctions. "The 'SLOSS' debate," Lovejoy would say, "in one sense was about the applicability of island biogeographic theory to reserve design, and it flourished as a controversy because there was so little direct data. Consequently I designed the project to do that"—to collect data, that is.

As Wilson and MacArthur and Wilson and Simberloff had shown, there was an inverse relationship between the area an animal population occupied and its extinction rate. All populations varied, increasing or declining according to available resources on the one hand and to challenges, from climate and weather to predation, on the other. If the area was smaller, the risk that one or an accumulation of those challenges might eliminate the entire population was real. It happened sometimes in rain-forest fragments where organisms that concentrated in a small area were entirely wiped out by a fire or a storm.

"Fragment size also influences the rate of species losses," one of the project's reports explains, "with smaller fragments losing species more quickly. Assuming the surrounding matrix is hostile to bird movements and precludes colonization, [one study] estimated that a 1,000-fold increase in fragment area would be needed to slow the rate of local species extinctions by 10-fold. Even a fragment of 10,000 [hectares] in area [about forty square miles, twice the size of Manhattan], would be expected to lose a substantial part of its bird fauna within one century."

Yet the rain-forest reserves couldn't occupy the entire country of Brazil; humans had a rightful place there as well. Some smaller part or parts of the vast Amazon basin and its lowland rain forest had to be reserved. Wilson summed up the conflict succinctly in *Biophilia* with a vivid image: "The action [of scraping away the rain forest for pasture or cropland] can be defended (with difficulty) on economic grounds, but it is like burning a Renaissance painting to cook dinner."

Lovejoy's research, and the alliances it promoted across the next three decades, led to significant rain-forest preservation. "The Brazilian Amazon," he commented in 2020, "despite all the deforestation and burning, has gone from two national forests to what . . . will amount to more than 40 percent receiving some form of protection." Yet destruction continues worldwide, an increasing challenge exacerbated in the present century by intensifying global heating.

Out of his Brazilian experience Wilson found the answer to the question of what he could do. He could do what he'd already begun doing: lecture and teach and write and testify. Shortly after publishing *Biophilia,* he wrote an essay on the biodiversity crisis meant for *Foreign Affairs,* a journal widely read among government officials. The editors rejected the essay as outside their range of subject matter, even though environmental damage powerfully affects the economies of developing countries. The journal *BioScience,* which features articles on public policy as well as research findings, welcomed the essay instead. "The Biological Diversity Crisis," published in December 1985, recalls Wilson's Brazilian rain-forest vision of spending years identifying all the species by name and learning the details of their lives. "About 1.7 million species have been formally named since Linnaeus inaugurated the binomial system in 1763," it reports. About 440,000 are plants, Wilson continues, 47,000 vertebrates, and 751,000 insects. But the "true magnitude" of life's diversity is in fact still a mystery, Wilson cautions. Citing several references, he reports estimates of three million insect species and, counting life in the oceans, perhaps ten million species overall.

Those numbers stood until 1982, when an entomologist at the National Museum of Natural History, Terry Erwin, and his colleagues developed a method for sampling the high canopy of tropical rain forests, which was "largely inaccessible," Wilson writes, "because of its height (a hundred feet or more), the slick surface of the trunks, and the swarms of stinging ants and wasps that break forth at all levels." Instead of climbing through that no-man's-land, the scientists fired a projectile with a line attached over an upper branch, hauled up a radio-controlled canister of insecticide harmless to vertebrates, laid sheets on the ground, released a chemical fog in the canopy, and collected the fallen prey. "Erwin," Wilson reports, "extrapolated a possible total of 30 million species" worldwide from the numbers of species he and his colleagues collected, "mostly confined to the rainforest canopy." (A 2018 re-examination using new data and improved statistical tools halves Erwin's estimate to about fourteen million, but notes that, "with 1 million insect species named, this suggests that 80% remain to be discovered.")

How had so many and such a variety of creatures been missed? The reasons, Wilson proposes, were both geography and "the natural human affections for big organisms. The great majority of kinds of organisms everywhere in the world are not only tropical, but also inconspicuous invertebrates such as insects, crustaceans, mites, and nematodes." Merely twenty-five acres of Borneo rain forest (about eleven city blocks) sustains as many species of trees—about seven hundred—as all the native species of trees in North America.

Such comparisons are interesting to those who collect odd facts, but, otherwise, why should we care? What practical value of biodiversity takes us beyond J.B.S. Haldane's well-known (and probably apocryphal) comment that the Creator, in making such an immense variety of them, demonstrated "an inordinate fondness for beetles"?

In his *BioScience* essay, Wilson framed his answer in terms of information. A species, he argued, is not like "a molecule in a cloud of molecules." It's a unique population of organisms, "the terminus of a lineage that split off thousands or even millions of years ago." Technically, it's "richer in information than a Caravaggio painting, Bach fugue, or any other great work of art." To illustrate what he meant, Wilson described the information density of a single strand of mouse DNA. Stretched out, such a strand would be about one meter (3.3 feet) long, but invisible to the naked eye (and small enough to curl up inside an equally invisible cell) because it's only twenty angstroms—two-billionths of a meter—in diameter. Yet "the full information contained therein, if translated into ordinary-sized printed letters, would just about fill all 15 editions of the *Encyclopaedia Britannica* published since 1768."

Drawing on his earliest scientific work of taxonomy, the work he did in graduate studies at Harvard, on his South Pacific expedition, and in Alabama before that, the work Jim Watson had denigrated as merely stamp collecting, Wilson first emphasized the importance to science of identifying and classifying all those millions of species. Doing so, he argued, could answer questions such as why thirty million (as he thought then) evolved rather than forty million or two thousand or a billion; or "why is there an overwhelming preponderance of insect species on the land, but virtually none of these organisms in the sea?" Current knowledge of the forces involved in the evolution of diversity, he concluded, is at about the level of physics as it was in the late nineteenth century—that is, before the discovery of radioactivity, before relativity or the first understanding of quantum phenomena, before the universe was discovered to be vastly larger than our home galaxy, the Milky Way. Great sections of the Earth's surface and most of the unique organisms that live on and in them remain *terra incognita* even as we obliterate them with fire and chain saw and bulldozer.

Not everyone values knowledge for its own sake, although the first step toward any practical use is surely identifying what something is made of, what it does, and how it works. Wilson used the example of food plants to demonstrate how limited our practical harvest has been across the ages:

> Only a tiny fraction of species with potential economic importance has been used.... A far larger number, tens of thousands of plants and millions of animals, have never even been studied well enough to assess their potential. Throughout history, for example, a total of 7000 kinds of plants have been grown or collected as food. Of these, 20 species supply 90% of the world's food

> and just 3—wheat, maize, and rice—constitute about half. . . . Yet waiting in the wings are tens of thousands of edible species, many demonstrably superior to those already in use.

Such was the promise of biodiversity. Then Wilson presented the shocking reality as of 1985. The forests of Madagascar had been reduced to less than 10 percent of their original cover. The Brazilian Atlantic forests were down to less than 1 percent. Within a century, eighty-four of the seven hundred species of birds in the Amazon basin would be extinct, as would three out of every twenty plant species in Central and South America. The current rate of extinction was four hundred times what it was before human civilization developed and was rapidly accelerating. The reduction in diversity was approaching the disaster of the sixty-five-million-years-past asteroid impact event that ended the age of the dinosaurs. "And in at least one respect," Wilson concluded his dark litany, "this human-made hecatomb is worse than any time in the geological past. In the earlier mass extinctions . . . most of the plant diversity survived. Now, for the first time, it is being mostly destroyed."

What could be done to stop this hecatomb—Wilson chose the old word for a great Roman sacrifice of one hundred oxen, which has come to mean a sacrifice of many victims—before biodiversity, with all its potential of knowledge and practical benefit, was bereft? The second half of Wilson's essay revealed its purpose as an indirect communication to the U.S. government. As he counted insects before, now he counted taxonomists and systematists—those scholars who identify, name, and catalogue species—and found their number nearly as depleted as that of the insects.

Just one thousand annual grants of fifty thousand dollars each, he proposed—fifty million dollars—would double the level of existing support for basic tropical biology, including systematics (identifying, naming, and describing organisms) and ecology (studying the relationships between organisms and their environments). That amount would be only 1.4 percent of the 3.5 billion dollars currently spent in the United States on health-related biology, he noted, and, slyly, "approximately equals the lifetime cost of one F15 Eagle fighter-bomber."

What was the United States' interest in these matters? "The problems of Third World countries, most of which are in the tropics, are primarily biological. They include excessive population growth, depletion of soil nutrients, deforestation, and the decline of genetic diversity in crop and forest reserves." Between 1975 and 1985, Wilson elaborated—in the previous ten years—virtually all official assessments had agreed "that the intricate economic and social problems of tropical countries"—he mentions Haiti, El Salvador, and Grenada as recent trouble spots—"cannot be solved without a more detailed knowledge of the environment." A timely report to the U.S. Congress on the conservation of biodiversity by an interagency task force that included the Agency for International Development, the Smithsonian Institution, and the Environmental Protection Agency allowed him to cite its "call for the primary inventory and assessment of native floras and faunas. In fact," Wilson added, "not much else can be accomplished without this detailed information."

He concluded his essay with a strong appeal for action:

> To put the matter as concisely as possible, biological diversity is unique in the evenness of its importance to both developed and developing countries and in the cost-effectiveness of its study. . . . This being the only living world we are ever likely to know, let us join to make the most of it.

But those were the years of Ronald Reagan's conservative presidency, when Reagan and others excoriated government as a cause of problems rather than a means of solving them. Wilson's proposal elicited no financial support. He would have to find another route to increasing the study of biodiversity and the protection of what remained in the world of wilderness.

12

Reprising Linnaeus

A PREVIEW OF that other route to identifying the millions of species yet unknown and unprotected appears in Ed Wilson's contribution to a national forum on biodivcrsity. Thc Scp-tember 1986 forum, Wilson reports, sponsored by the National Academy of Sciences and the Smithsonian Institution, "featured more than 60 leading biologists, economists, agricultural experts, philosophers, representatives of assistance and lending agencies, and other professionals." Its lectures and panels drew large audiences and national press coverage. Wilson thought the event signaled a noticeable rise in interest in both biodiversity and international conservation, for two reasons: enough study had been devoted across the decade to "deforestation, species extinction, and tropical biology to . . . warrant broader public exposure"; and developing nations in particular had become aware of "the close linkage between the conservation of biodiversity and economic development."

Wilson's insight turns up in the paper he developed from his keynote address, "The Current State of Biological Diversity." There he points out the shocking fact that "the number of species [on Earth] is not known, even to the nearest order of magnitude," while species are going extinct at the rate of roughly one a day. If species are vanishing, so are scientists of the kind trained to identify and study them: "Probably no more than 1,500 professional systematists in the world are competent to deal with the millions of species found in the humid tropic forests" alone, Wilson said, and the number of such specialists was declining "due to decreased professional opportunities, reduced funding for research, and the assignment of a higher priority to other disciplines."

There is no hope of saving species from decline and extinction unless we know which they are, where they live, what their biology is, and how vulnerable they are to environmental change. Therefore—and here was the preview of what would become for Wilson the central project of his life immediately post-Harvard—

> It would be a great advantage, in my opinion, to seek such knowledge for the entire biota of the world. Each species is unique and intrinsically valuable. We cannot expect to answer the important questions of ecology and other branches of evolutionary biology, much less preserve diversity with any efficiency, by studying only a subset of the extant species.

Then, as if he himself had not grasped the full scale of his proposal until the moment of writing those words, he follows out its challenge and its consequences:

> I will go further: the magnitude and control of biological diversity is not just a central problem of evolutionary biology; it is one of the key problems of science as a whole. At present, there is no way of knowing whether there are 5, 10, or 30 million species on Earth. There is no theory that can predict what this number might turn out to be. . . . Unless an effort is made to understand all of diversity, we will fall short of understanding life in these important respects, and due to the accelerating extinction of species, much of our opportunity will slip away forever.

When I pointed out this 1986 reference to Wilson, it surprised him; he had forgotten how early he had begun thinking about identifying every individual species on Earth. In a real sense, he agreed, that interest extended all the way back to childhood; he had always been captivated by finding, identifying, and naming what he sometimes called "the little things that run the world"—and the big things, too.

He and Peter Raven, the activist director of the Missouri Botanical Garden, had shared long discussions during the 1990s about how to make such an ambitious project happen. They were hampered by the limited capabilities of the digital media of the day; it's easy to forget that Tim Berners-Lee conceived of the World Wide Web only in March 1989, and that the first Web page was served on the open Internet only at the end of 1990. In November 1992, Wilson and Raven published a proposal in the journal *Science* for a fifty-year plan for biodiversity surveys. They wrote of "the roughly 170,000 flowering plants and 30,000 vertebrates" then known and described, of another 250,000 that "appear[ed] to have been described thus far," but cited estimates of the remainder of undescribed species "ranging from 8 million to 100 million" to emphasize the magnitude of the task at hand. This "great problem," they argued, required "a wholly new approach." They proposed combining each nation's national surveys to build a world inventory, but digital approaches escaped mention.

Wilson remembers the 1990s as a time when he and a few others were voices crying in the wilderness, "wandering around, saying, 'You know, we need to bulk up the exploration of the planet, the living part,' and trying to raise a lot of money for it, unsuccessfully." In 1995, when the U.S. Congress debated renewing the 1973 Endangered Species Act, Wilson joined his old friend Tom Eisner and other scientists and environmental activists in lobbying for the act's renewal while arguing for the continued protection of individual species rather than shifting to broader but coarser ecosystem-level protection. "Each species," they wrote, "by virtue of its genetic uniqueness, is the source of information we can learn from no other source." Ironically, the argument their opponents presented for shifting away from protecting individual species was "the lack of ecological information about most species."

At the beginning of the new century, others joined Raven and Wilson in looking for a way to catalogue the Earth's immense variety of life. Stewart Brand, the creator of the *Whole Earth Catalog* (and, more recently, with the inventor and entrepreneur Danny Hills, the Long Now Foundation), recalls a dinner at the house of Microsoft's chief technology officer, Nathan Myhrvold, when a number of wealthy guests complained about how hard it was to give money away responsibly and lamented the absence of a single grand project they could contribute to. One of the dinner guests, Kevin Kelly, a photojournalist who had cofounded *Wired* magazine in 1993 and served as its executive editor for the next seven years, mused about the possibility of identifying every species on Earth. If we landed on a new planet, Brand remembers Kelly saying, that's one of the first things we'd do—identify all the life-forms—but we haven't done that yet for our own planet.

"I was taken by the idea," Brand told me. "That was in 1999, when money for imaginative projects was falling out of the trees, especially if they related to the Internet." Brand, his wife and business partner Ryan Phelan, and Kelly organized a conference at the California Academy of Sciences in San Francisco in September 2000 to discuss establishing an All Species Foundation "to catalogue every living species on earth within one human generation (twenty-five years)." Phelan and Brand had sought out Wilson for advice on the project; seeing its connection to his own efforts, he had immediately offered his help. Raven and Wilson subsequently served as scientific advisers to the new foundation. Phelan, a longtime entrepreneur and consultant, directed it. It began operation in 2000 with a founding grant of one million dollars.

"For a good year," Phelan told me, "Ed was my go-to person, warm and courtly and always helpful. One of the reasons I loved running All Species was the opportunity I had to work with Ed. The big problem was assembling data. A lot of databases were blocked off behind either a paywall or an institutional wall. We decided to hack them.

We weren't stealing, just using the information to prepare a demonstration. We were going to set up a separate page for each life-form. So we hacked the information and made up Web pages and showed them to the scientists." Stewart Brand continues: "The imaginative ones were delighted with what we were trying to do. The stodgy ones were extremely hostile, as in 'Don't fuck with my data!' "

The next year, 2001, with the economic downturn that followed the 9/11 terrorist attack on the World Trade Center, the trees stopped shedding grant money and the foundation faltered. "We all hoped that, working together, we could pull off the grand vision," Phelan concludes. "The fact that we couldn't fund the project the way we hoped was a great disappointment—but Ed took it to the next stage, and that was a consolation."

Wilson's ideas for a universal catalogue of species coalesced across the next months into a bold proposal he called "the Encyclopedia of Life." He abstracted the proposal in a scientific journal in February 2003:

> Comparative biology, crossing the digital divide, has begun a still largely unheralded revolution: the exploration and analysis of biodiversity at a vastly accelerated pace. Its momentum will return systematics from its long sojourn at the margin and back into the mainstream of science. Its principal achievement will be a single-portal electronic encyclopedia of life.

"At a deeper level," Wilson continued in the body of the proposal, "such an encyclopedia would transform the very nature of biology, because biology is primarily a descriptive science":

> Each species is a small universe in itself, from its genetic code to its anatomy, behavior, life cycle and environmental role, a self-perpetuating system created during an almost unimaginably complicated evolutionary history. Each species merits careers of scientific study and celebration by historians and poets. Nothing of the kind can be said (at the risk of stating the obvious) for each proton or inorganic molecule.

Wilson's old battle with the molecular biologists echoes in that final sentence.

"What we need is to get out there and search this little-known planet," Wilson recalls explaining his idea, "and then put all the information into a single great database, an electronic encyclopedia, with a page that's indefinitely extensible for each species in turn, and that would be available to anybody, anytime, anywhere, single access, on command, free."

This time around, the idea caught fire. Jonathan Fanton, an Alabama-born educator who had been president for many years of Manhattan's New School for Social Research, was appointed president of the John D. and Catherine T. MacArthur Foundation in 1999 and was looking for large, signature projects to support. "That was the key," Wilson says. The MacArthur and Alfred P. Sloan Foundations eventually contributed some twelve million dollars to launch the Encyclopedia of Life.

Wilson officially announced the project in a TED Prize talk he gave in Monterey, California, in 2007. The TED Prize, which was then worth a hundred thousand dollars, was awarded annually beginning in 2005 to three recipients chosen for the appeal of their stated "wish to change the world." Besides Wilson, former president Bill Clinton and the photojournalist James Nachtwey received 2007 awards.

Wilson spoke at length about his long-standing interest in identifying species. "We live on a mostly unexplored planet," he told the TED audience provocatively. "The great majority of organisms on Earth remain unknown to science." Two new kinds of whale had been discovered recently, he said, two new antelope, dozens of species of monkey, a new kind of elephant, and even a distinct kind of gorilla. "At the extreme opposite end of the size scale, the class of marine bacteria, the *Prochlorococci* ["that will be on the final exam," he quipped], although discovered only in 1988, are now recognized as likely the most abundant organisms on earth." And these, he said, are only a first glimpse of "our ignorance of life on this planet."

Worse, Wilson continued, our knowledge of biodiversity is so incomplete "that we are at risk of losing a great deal of it before it is even discovered." For example, he said, the two hundred thousand species currently known in the United States were actually only partially understood: "Only about 15 percent of the known species had been studied well enough to evaluate their status," and of those evaluated, "20 percent are classified as 'in peril,' that is, in danger of extinction." And that was only in the United States. "We are, in short, flying blind into our environmental future.... We need to settle down before we wreck the planet."

These facts and arguments led Wilson directly into his TED Prize wish:

> This should be a big science project equivalent to the Human Genome Project. It should be thought of as a biological moonshot with a timetable. So this brings me to my wish for TEDsters, and for anyone else around the world who hears this talk: *I wish that we will work together to help create the key tool that we need to inspire preservation of Earth's biodiversity: the Encyclopedia of Life.*

He elaborated on his wish in a description for the TED Web site:

> The Plan: Create an online encyclopedia with a webpage for every species containing articles, photos, maps, data and links to communities. Develop a platform where users can find trusted, consolidated information from resources around the world. Become a critical component of the global bioinformatics infrastructure for researchers, citizen scientists and bio-friendly communities.

The five original cornerstone institutions for the Encyclopedia of Life, with an eventual fifty-million-dollar funding commitment, included Chicago's Field Museum of Natural History, Harvard University, the Marine Biological Laboratory at Woods Hole, the Missouri Botanical Garden, and the Smithsonian Institution. The site (www.eol.org) went live in February 2008. Wilson calls it "a dream come true."

All the work Wilson had done in the post-*Sociobiology* years, both scientific and popular, especially the books he had written, had brought him to wide public attention. Marking the beginning of this phase of his long life were two major awards he received, both jointly, at the beginning of the 1990s. The first was the Crafoord Prize, a Swedish prize set up by the family whose medical-instrument firm, Gambro AB, manufactures devices sold worldwide for liver and kidney dialysis. Awarded in partnership with the Royal Swedish Academy of Sciences, the six-million-Swedish-krona prize (about $720,000 today) honors disciplines that the Nobel Prizes omit, including astronomy and mathematics, the geosciences, and the biosciences. Wilson won the Crafoord, divided with the Stanford University entomologist Paul R. Ehrlich, "for the theory of island biogeography and other research on species diversity and community dynamics on islands and in other habitats with differing degrees of isolation."

The second award Wilson received was a second Pulitzer Prize, awarded jointly with a German colleague and good friend, Bert Hölldobler, for their 1990 book *The Ants*. Hölldobler, a gifted myrmecologist, had been a professor of biology at Harvard since 1973. He was an experimentalist rather than a field biologist; his laboratory work complemented Wilson's field studies, and the two men meshed well. But Hölldobler had difficulty raising the funding he needed to support his American laboratory, and in 1989 the University of Würzburg recruited him to assume the chair of behavioral physiology and sociobiology at its Theodor Boveri Institute, with a million-dollar commitment.

"Our collaboration had been very fruitful," Wilson remembers. "We had different talents and different approaches. We published a lot. So before he left, we said, 'Look, between the two of us, we know everything there is to know about ants. Let's write a book and put it all down.'" The book they produced was vintage Wilson in form, a big, well-illustrated coffee-table-sized compendium of 732 glossy pages. It was a work of serious science, however, just as the similarly formatted *The Insect Societies* and *Sociobiology* had been; Wilson was astonished that it won a purely literary prize. I was not, then or now, because I was the member of the nominating jury who recommended it to the Pulitzer Prize board.

I knew Wilson slightly at that time. I had interviewed him ten years previously for a magazine article I was writing. I remembered the encounter vividly because he'd taken me down the hall from his office at the Harvard Museum of Comparative Zoology to show me the indefatigable Alfred Kinsey's collection of more than five million gall wasps, preserved on pins in drawer after drawer of a roomful of entomology storage cabinets, the subject of his Harvard Ph.D. dissertation. (Gall wasps stimulate trees and plants to grow wartlike swellings in which wasp larvae can develop safely while feeding on plant juices. Kinsey had collected gall wasps before he began collecting data on human sexuality; some wags joked that he'd used the same techniques in both cases.)

A jury of three nominating judges in each Pulitzer book category scours hundreds of books to find three finalists to recommend to the Pulitzer Prize board, which then selects the winner. Robert O. Wyatt, the book editor of the Nashville newspaper *The Tennessean* and a journalism professor, was chair of the 1991 nominating jury; the third judge was the writer Annie Dillard. Shipments of nominated books had begun arriving at our homes the previous summer—dozens of books, hundreds of books. It was impossible to read them all, of course, but it was possible to eliminate many of them—e.g., routine cookbooks and illustrated books on home repair—with a quick skim. (I was puzzled that such books had been submitted in the first place, but the submission fee wasn't high, and perhaps their authors' egos needed soothing. Is it better to have your book submitted and left behind at the starting gate or not to be submitted at all?)

I read *The Ants* with increasing appreciation. Of Wilson's work I had by then read only *On Human Nature* and *Sociobiology.* I knew what a good—and stimulating—writer he was. This time around, he and Hölldobler had accomplished something extraordinary: they had written a work of formal science that was also a distinguished work of literature. A Pulitzer Prize for such a book seemed to me a fine idea. I sent my proposed selection to the other two judges; they sent me theirs. I read both, but thought neither of them matched *The Ants* in depth or breadth.

We spoke by phone. Each judge had a favorite and lobbied hard for it. None of us were willing to rank our choice other than first. The Pulitzer board required a half-page recommendation for each finalist. Someone suggested we each write the recommendation for our favorite book and submit them unranked. I agreed immediately; doing so meant I had a chance to write directly to the board about why I thought it should award a Pulitzer Prize to a formal work of science.

The strategy succeeded. The Pulitzer board chose *The Ants* for its 1991 Pulitzer Prize in general nonfiction. It was Wilson's second Pulitzer and, as far as I was concerned, well deserved. The day the prizes were awarded, in the Low Library rotunda at Columbia University, my wife and I and several friends took Wilson and Hölldobler to a celebratory late lunch at a midtown restaurant. It was a merry party. Wilson had to leave before dessert, to get back to Boston and Irene. He told an interviewer in 2014 that the prize had astonished him. "We didn't write [*The Ants*] for the literary world," he said. "It was the only book of science—meant for scientists—ever to win the prize." That works of science can also be, and sometimes are, works of depth and grace, was exactly the point. Darwin's *On the Origin of Species* was such a book, a worthy model and predecessor. As Wilson would write, with a colleague, in an essay on scientific natural history, "The story of any species chosen at random is an epic, filled with mysteries and surprises that will engage biologists for generations to come." And how much more so the story of an entire genus representing thousands of species?

The day after Wilson learned he would be awarded his second Pulitzer Prize, he happened to be speaking at the 1,727th stated meeting of the venerable American Academy of Arts and Sciences, located conveniently in Cambridge, Massachusetts. The title of his talk was "Ants." He might have spoken about "a wide-ranging subject such as the relation of biology to social sciences or the global environmental crisis," he teased the assembled worthies, but he "saw instead an opportunity of truly historic proportion—namely, to give a talk with the shortest title in the history of the Academy."

Having thus lightly punctured the two-century-old academy's intellectual pretensions, Wilson proceeded to introduce his subject:

> The question I'm asked most often about ants is, "What do I do about the ones in my kitchen?" And my answer is always the same: "Watch where you step. Be careful of little lives. Feed them crumbs of coffeecake. They also like bits of tuna and whipped cream. Get a magnifying glass. Watch them closely. And you will be as close as any person may ever come to seeing social life as it might evolve on another planet."

In the course of his talk, Wilson mentioned that he was "currently studying" a genus of ants, *Pheidole,* that contained some 250 species already known to science—that is, named and described—from the New World alone. He had about six hundred species in the MCZ collection at Harvard, he said, which meant that some 350 in the collection were new to science. "More pour in from collectors every few months."

He himself, he said, had recently discovered a new species of *Pheidole* "in the office of the president of the World Wildlife Fund. . . . It was in a potted plant, this new species. As a member of the Board of Directors of the WWF, I forbade spraying the colony because it was, at least at that time, the only known living population of the species." Having brightened the academy chambers with this startling story of aliens living among us, Wilson went on to offer the assembled a friendly primer of ant biology.

Studying *Pheidole,* as it turned out, was one way Wilson recentered himself after the years of controversy over sociobiology. Not only were there undescribed specimens of the genus languishing at the MCZ; even more specimens languished in similar museum collections throughout the United States and across the world. *Pheidole* was the biggest and most complex undescribed genus of all the ants.

Wilson had always had a special interest in it, he told me, because of its wide distribution—a sign of its evolutionary success—and because it displayed "lots of interesting behavior."

"When I formally retired from Harvard in 1996," Wilson said, "I wanted to plunge back into the study of ants. I wanted to confront a major challenge and solve it. I might have preferred to go on a collecting expedition, but I wasn't in a position to do so at that time. So, instead of fishing in a rain forest, I went fishing in the MCZ collection." Eventually, he would extend his fishing to other museums and in correspondence with other myrmecologists, but he worked the MCZ collection first. *Pheidole* are small ants. He would load up a batch of unidentified specimens, take them home, and work with them one by one under a microscope in his home laboratory, cataloguing features, making measurements, doing drawings. "It was a real pleasure," he told me. "I got to stay at home and listen to music." For the book-length catalogue he was planning, a labor of love titled *Pheidole in the New World,* he would delineate every species with line drawings of its major and minor forms. Before he was finished, he had completed some five thousand drawings. "Now, that might sound rather odd," he told an interviewer, "but I find it particularly rewarding. I'm doing that on the side. That's sort of like a hobby." Below each set of drawings he presents a formal description of the specimen, such as:

> DIAGNOSIS: A small, light-colored member of the *flavens* group similar to the species listed in the heading above, distinguished in the **major** by the bulbous pronotum in dorsal-oblique view, set off from a small but distinct mesonotal convexity, and the cephalic rugoreticulum, which starts as a patch at each occiput corner and runs anteriorly in a thin band to a patch just mesad to the eye. The **minor** is distinctive in the steep, nearly vertical descent of the posterior mesonotal face in the metanotum.

The ant's measurements followed, its color, its range, its biology ("The type series was collected on a steep, rocky forested slope [(by) W. L. Brown]; the Belize ants were taken from the ground in a Cecropia-palm forest"). To this particular species, new to science and thus described for the first time, Wilson awarded a distinctive species name: *Pheidole harrisonfordi.* "Named in honor of [the actor] Harrison Ford," Wilson noted in the "Entomology" category under the drawings, "in recognition of his outstanding contribution in service and support to tropical conservation, hence the habitats in which the Pheidole ants will continue to exist." Ford had contributed in support of the E. O. Wilson Biodiversity Foundation, a nonprofit established in Durham, North Carolina, in 2005 to honor Wilson's work through biodiversity research and education. One of its programs was support for the Encyclopedia of Life.

All this information was arranged on a single page if possible, continuing to the next page if necessary. In this manner, the catalogue worked its way through the entire *Pheidole* genus—as much of it as Wilson could find and identify, a total of 624 species—presented in alphabetical order in a heavy coffee-table volume of 794 pages.

Wilson had kept up with the development of sociobiology during his final Harvard years and afterward. By 2000, though his old adversaries continued their attacks (Gould would die of cancer in 2002), one major scientific database identified more than thirteen thousand entries under the word "sociobiology." Five major journals reported on developments in the field, and sociobiology research and teaching reached worldwide. Critics who claimed Wilson had done nothing new, writes Ullica Segerstråle, had merely assembled a mass of information, "perhaps fail to recognize the *kind* of contribution Wilson made: he created a field by demonstrating to its potential members that it existed—partially by co-opting them as contributors to his project!"

Wilson had built sociobiology on the foundation concept of the selfish gene—Bill Hamilton's 1964 inclusive-fitness model that explained "unselfish" altruism on the basis of selfish kin selection. Members of a society, that is, were thought to act at a disadvantage to themselves when in fact they increased the odds that related members would pass along the genes they had in common, and the more closely they were related—the more genes they had in common—the greater the odds of seemingly altruistic sacrifice.

The Hamilton model found favor in the years after Hamilton's 1964 paper largely because it seemed to explain the eusocial insects so well. Among such insects—ants, bees, wasps, termites—the paradoxical willingness of sterile female workers to sacrifice their lives raising the queen's offspring had challenged Darwin almost to the point of defeating his grand idea of natural selection. "One special difficulty," he reported in the *Origin of Species*, ". . . at first appeared to me insuperable, and actually fatal to my whole theory. I allude to the neuters or sterile females in insect communities: for these neuters often differ widely in instinct and in structure from both the males and fertile females, and yet from being sterile they cannot propagate their kind." Darwin explained this paradox by proposing that selection operated at more than one level. Within a group, selfish individuals might outcompete altruists; but an altruistic group would outcompete a selfish group. As long as the balance favored group competition over individual competition, the group could survive to propagate its members' altruistic traits along with their selfish ones.

Rather than welcome Darwin's insight, however, biologists followed Hamilton's preferred kin selection ("helping one's own genes in the bodies of others," Wilson defines it), or Bob Trivers's 1971 reciprocity argument ("helping others in expectation of return benefits"). Both kept the selection process within the same level of individual advantage: I'm risking myself by helping you because we're related and you'll pass on some of my genes; or I'm risking myself by helping you because I expect you to help me in return, increasing my chances of reproducing. Between-group selection was assumed to be delicately balanced against within-group selection, an assumption, Wilson says, made "primarily to simplify the mathematics" at a time when computing power was limited.

For these and other reasons, group selection was widely rejected among biologists in the final decades of the twentieth century. Even Wilson went along with Hamilton and Trivers in their dismissal of group selection and its replacement with selfish-gene models, despite the problems those models posed for sociobiology.

Yet there was never any large body of evidence supporting the rejection of group selection, Wilson reports. Rather than finding evidence to demonstrate its supposed weaknesses, critics argued against its plausibility; their arguments were mostly theoretical. Out in the field, in contrast, Wilson argues, social vertebrates such as lions "provide convincing evidence for group selection."

Why was the question of group versus individual selection important? Because it concerned what Wilson calls "the fundamental problem of social life," which is that selfishness defeats altruism within a group, and therefore altruism ought to decline to extinction, yet it persists, and is in fact the very basis of social organization at every level, from insects to humans.

Several lines of evidence nagged at Wilson in the 1990s and the early 2000s. One was the relative rarity of eusociality despite its signal success in those few species where it occurred: of about twenty-six hundred living taxonomic families of insects and other arthropods recognized at that time, Wilson writes, only fifteen were known to contain eusocial species. Another was the lack of evidence that relatedness was even necessary for eusociality to evolve. Finally, Hamilton's hypothesis of haplodiploid reproduction in eusocial insects—of sisters, that is, sharing three-fourths of their genes whereas males share only one half, making sisters genetically closer to one another ("super-sisters") than they would be to their own offspring, and thus predisposed to support their sister queen's offspring without reproducing themselves—had collapsed in recent years: haplodiploidy and eusociality had been found to have evolved separately frequently enough to be considered separate processes.

To most biologists, the selfish-gene model had seemed a major step toward completing the revision and extension of Darwinism necessary to round out the Modern Synthesis; by the turn of the twenty-first century, it was widely regarded as the correct, reigning paradigm. Wilson had already proved himself to be no respecter of paradigms when the evidence seemed to point in another direction. In 2005, with Bert Hölldobler, he published a paper in the *Proceedings of the National Academy of Sciences* rejecting the selfish-gene model outright as the basis for the development of eusociality. "Eusociality: Origin and Consequences" seems not to have aroused much opposition in the biology community, but a paper Wilson published in 2010 in *Nature,* "The Evolution of Eusociality," written with two young Harvard colleagues, the mathematical biologists Martin A. Nowak and Corina Tarnita, started another fierce sociobiology war, still ongoing ten years later.

A British philosopher of science reviewing the debate, Jonathan Birch, calls the *Nature* paper "incendiary"; it presented, he writes, "a savage critique of . . . W. D. Hamilton's theory of kin selection. More than a hundred biologists [137 in one *Nature* response alone] have since rallied to the theory's defense, but Nowak et al. maintain that their arguments 'stand unrefuted.'" Further rejections and attempted refutations have followed in the years since then, including by such well-known figures as Richard Dawkins and Steven Pinker, who had previously been Wilson's bulldogs.

According to Nowak, Tarnita, a vivacious young Romanian woman who was a Junior Fellow at the time their paper was published, "sought out people on the other side of the debate to discuss the work, and they were disarmed. They tried to find mistakes. There was no mistake."

Wilson enjoyed working with the two mathematicians. Tarnita, now an associate professor at Princeton, "is a mathematician of the first rank," he told a journalist in 2011. Wilson keeps a bobblehead doll of Charles Darwin on his desk in the MCZ; when the team worked together in his office, he liked to ask the wee Darwin if the great scientist agreed with their conclusions. Darwin bobbled enthusiastically, Wilson reports: "We got a strong affirmative."

The American economist and mathematician Herbert Gintis, reviewing Wilson's book *The Social Conquest of Earth* in 2012, reports that Wilson's rejection of Hamilton's model "has shaken the very foundations of population biology." Wilson remains confident that his argument is valid, not least because his two younger colleagues got the math right. Even Hamilton eventually realized that kin selection is in fact a kind of group selection, Wilson says. "We hope our new theory for the evolution of eusociality will open up sociobiology to new avenues of research," he told *The Harvard Gazette*. ". . . After four decades ruling the roost, it's time to recognize [kin selection] theory's very limited prowess."

Most vividly, Wilson and an unrelated colleague, David Sloan Wilson, summarized the evolving new model of the evolution of eusociality in a 2007 paper exploring the issues that would roil biology in 2010. "When Rabbi Hillel was asked to explain the *Torah* in the time that he could stand on one foot," they wrote there, "he famously replied: 'Do not do unto others that which is repugnant to you. Everything else is commentary.' Darwin's original insight and the developments reviewed in this article enable us to offer the following one-foot summary of sociobiology's new theoretical foundation: 'Selfishness beats altruism within groups. Altruistic groups beat selfish groups. Everything else is commentary.'"

A different and, for Wilson, delightful opportunity came his way in 2011, when an Idaho native, a former entrepreneur turned philanthropist named Greg Carr, invited him to visit a fifteen-hundred-square-mile national park Carr was restoring in the Southeast African country of Mozambique. Carr had committed himself to full-time philanthropy in the late 1990s, after creating several early Internet companies that had made him wealthy. Carr sought hands-on commitment, and when the Mozambican ambassador to the United Nations invited him to come to his country and help it—Mozambique was just beginning to recover from a long, destructive civil war—Carr had discovered Gorongosa National Park on a helicopter tour of the country and made restoring it his personal cause. In 2004, the government of Mozambique and the Carr Foundation had agreed jointly to restore the ruined park; Carr committed forty million dollars to the cause.

Inviting Wilson to visit in 2011 was part of Carr's program. It eventuated in 2014 in the opening of the E. O. Wilson Biodiversity

Laboratory there, offering long-term research and training in biodiversity documentation, ecology, and conservation biology to visiting researchers from both Mozambique and abroad. It may be the only such laboratory in the world located in a national park. Consistent with Carr's commitment, it incorporates jobs and science education for local-area populations.

Wilson visited Gorongosa again in 2012, this time leading an MCZ expedition to study the many species of previously unidentified ants in the park. At the same time, Wilson would participate in filming for the *Life on Earth* online high-school biology textbook then under production, another of his many efforts to expand biological knowledge across as large a range of humanity as possible. Among other experiences, the MCZ team witnessed a swarm of driver ants moving through their camp in its implacable hundreds of thousands, although this time the swarm was moving to a new nest rather than sweeping a thirty-foot fan of savanna completely clean of animal prey. Even standing near the outer edge of the flow, the team members suffered only a few stings from ants crawling up their ankles. Drawing on his 2012 visit, Wilson wrote a small but elegant book, *A Window on Eternity: A Biologist's Walk Through Gorongosa National Park,* illustrated with color photographs, published in 2014.

Wilson planned to visit Gorongosa again in April 2019 to spend a month doing fieldwork. I planned to join him there for several weeks as well. (He travels these days with Kathy Horton, who became his reliable travel associate in 2007 when a cataract had developed in his one good eye that made it difficult for him even to find his airplane seat. Successful cataract surgery has successfully cleared his eye, but Horton's assistance continues.) He had already reached South Africa, and I had received my inoculations and assembled my supplies, when a powerful cyclone (as hurricanes are called in that region of the world) struck Mozambique. It destroyed 90 percent of the small coastal city of Beira, including the international airport, flooded much of Gorongosa, and made travel impossible on the country's limited road system. The park closed, so that the park staff could use its helicopter and ground vehicles to ferry food to villages in the

Mozambique interior that were completely cut off from supplies. The COVID pandemic restricted travel in 2020, but Wilson, although now in his early nineties, planned to return to his window on eternity when next he could.

Wilson's engagement with Gorongosa, his high-school biology project, his *Encyclopedia of Life,* and his recent books, he told me, all serve the larger purpose he has made the focus of his work in the final decades of his life: to advance the cause of preserving biodiversity by producing "a complete account of the rest of life on earth." Such an effort, he believes, must encourage people to think seriously, "as a world culture," about the natural world of which human beings are themselves an integral part. As the world goes, so, ultimately, go we. If we destroy the world, we destroy ourselves. We will not escape on shining white rockets to some other haven. "How to serve both humanity and the rest of life," Wilson writes near the end of *A Window on Eternity,* "is the great challenge of the modern era."

Wilson's most recent response to that challenge is an ambitious project he calls "Half-Earth." As he describes it in a 2016 book of that title, Half-Earth proposes "that only by committing half the planet's surface to nature can we hope to save the immensity of life-forms that compose it. . . . The Half-Earth proposal offers a first, emergency solution commensurate with the magnitude of the problem." The half of the Earth Wilson has in mind is what remains of wilderness throughout the world—preserving the large plots and reconnecting the smaller plots with corridors. Once again, Wilson's previous work with Robert MacArthur on island ecology supplied a scientific basis for understanding the scale of wilderness plots required to prevent the extinction of species.

13

Origami

WRITING A BIOGRAPHY of a living person allows for interviewing him or her, something impossible with a subject who's deceased. When I began research for this book, in the autumn of 2018, before the COVID pandemic made travel dangerous, I regularly visited Ed Wilson at Brookhaven at Lexington, his Massachusetts retirement community. Ed and Irene had moved there in 2001—on 9/11, as it happened—to relieve themselves of the chores that went with owning a home. I stayed in a guest apartment on the premises, which made it easy to interview Ed for several hours in the morning, to break together for lunch, and then to work several more hours in the afternoon.

During one such visit, in November 2018, Ed told me he was expecting a guest for lunch the next day. I said I'd like to join them. He thought about it for a moment and agreed.

We met on that Monday morning in the front lobby of the residential apartment building where the Wilsons lived and adjourned to a garden room filled with green plants that looked out onto a courtyard behind the building. Busy with interviewing, we worked through the morning until the lunch hour. As I packed away my recorder, I asked Ed whom we were meeting for lunch.

"Paul Simon," he said. "Paul Simon and his brother, Ed."

"The singer?" I said, startled.

"Yes. Let's go."

We walked down the hall, turned a corner, walked down another hall. Along the way, Ed explained that Simon was a supporter of his Half-Earth Project to save what was left of wilderness in the world. "He donated the profits from a concert tour to Half-Earth," Ed said.

We came to a small private dining room. I opened the door for Ed and followed him in. The Simon brothers were already there, seated together at one side of a round dining table. I held back until Ed could introduce me, which gave me time to notice that the table was strewn with hearing aids, a dozen or more.

We went through introductions. Paul Simon was wearing a leather jacket. His signature fedora was pushed back on his head. His brother, who was Paul's height and bore a close brotherly resemblance, preferred a baseball cap. "We brought Edie up for her concert tonight," Paul told Ed Wilson. "We thought we'd stop off and see you." Edie was Paul Simon's wife of twenty-six years, the singer and songwriter Edie Brickell. The Simons lived in Connecticut. Edie Brickell and her group, the New Bohemians, were giving a concert that night in Boston. Paul gestured to the table. "You said you were having trouble with your hearing aids. We brought some new ones for you to try out."

"Why, that's very kind of you," Ed Wilson said, Alabama-polite.

"Ed's a good sound man," Paul Simon added, indicating his younger brother.

"Can I see your hearing aids?" Ed Simon asked Ed Wilson. Wilson removed them left and right and handed them over, little lima-bean halves with clear-coated wires that curved around to cone-shaped in-ear speakers.

Ed Simon studied them. "Model T's," he said.

Ed Wilson nodded. "I've had them for years."

"Lots of improvements since then," Ed Simon said.

It went like that for a while. A waiter came in—a student, dazzled but trying for casual. Paul Simon wanted simple scrambled eggs, not a problem. We ordered club-style, writing our orders down on little tickets with sharpened pencil stubs. The waiter left. Ed Simon worked through his collection of up-to-date hearing aids, one at a time, explaining their features, and Ed Wilson tried them on. I talked to Paul Simon. We got onto physics. Paul commented briefly but expertly on harmonics, how seldom they're mentioned but how important they are to song melodies. I asked him about his decision to retire—his concert series the previous summer, "Homeward Bound—The Farewell Tour." He said the worst part of touring was the time between concerts. "I can't go out to dinner with friends because I can't talk. I have to save my voice for the concerts. You end up stuck in your hotel room like a zombie." He was married, with children, and it was lonely. Nineteen U.S. cities between 16 May and 20 June. Nine cities across Europe between 30 June and 15 July. Years of that. Francis Bacon: "He that hath wife and children hath given hostages to fortune; for they are impediments to great enterprises, either of virtue or mischief." He hated it.

Our lunches arrived, with two more waiters helping carry to collect bragging rights. Ed Simon moved his cache of hearing aids to one side. Ed Wilson finished up trying them on. "None of them seem right," he said finally.

"Do you ever get down to New York?" Paul Simon asked. "We've got a really excellent audiologist down there. She can find one that works for you."

"Not recently," Ed Wilson said. "I don't travel that much these days." In 2018, he was eighty-nine. Certainly still vigorous, body and mind. The previous evening, on our way to dinner, I'd noticed the crowd of quad canes and four-wheeled rollators skulking outside the dining room, like a murder of crows. Ed had dismissed them. He disapproved of canes and walkers, he said. People got used to them and lost their sense of balance. Not his problem, even with only one functioning eye.

"Well, look," Paul Simon said. "Let's make it easy for you. I'll arrange for a car to take you to the airport. We'll fly you down to New

York. I'll have a car waiting for you. You can see our audiologist. Then we'll fly you back up here. I'll send my plane."

I'll send my plane.

Simon had attended Wilson's 2007 TED Prize talk, the one where Bill Clinton also spoke. He'd listened and been moved. He found Wilson afterward and told him he wanted to help. He recalled the moment a decade later in an essay for *The New York Times Book Review*'s "Year in Review" section. "I heard Wilson speak ten years ago at a TED conference," Simon wrote. "Something he said during his talk concerning our ecological problems stayed with me ever since. Its glistening optimism gave me hope. . . . Paraphrasing what he said: This planet has the potential to be a paradise by the next century if we act now to preserve it. A paradise!"

By the second decade of the twenty-first century, Ed Wilson had achieved world status as a scientist and an activist for biodiversity. People such as Paul Simon, Greg Carr, and the actor Glenn Close sought him out to offer their support. A group of medical-technology entrepreneurs established the Edward O. Wilson Biodiversity Foundation in Wilson's honor in 2005; it was to that organization that Paul Simon contributed his concert profits.

Wilson is the author of thirty-four books and more than 433 scientific papers. With *Sociobiology,* however controversial, he founded a new field of scientific research. He has received more than forty-five honorary degrees internationally and more than 150 awards and medals. He is a member, honorary member, or foreign member of more than thirty-five scientific organizations and societies, ranging from the Royal Society to the Worldwide Dragonfly Association to the Explorers Club. He's lectured across the world. All this while living quietly and continuing to explore the natural world as he finds it refracted through reports of revealing experiments and new discoveries.

I wrote this biography in part because I saw in Wilson a quality rare among human beings: he has never stopped growing in knowledge or expanding in range. When I began investigating Wilson's life,

I discovered I wasn't the first to think so. Nicholas Wade, the *New York Times* science writer, commented on Wilson's unusual capacity for growth in a profile published back in 1998:

> In an alternative fate, Dr. Wilson might have been an obscure expert on the ants of Alabama, his home state. But at each stage of his career he has looked outward, trying to see how the scholarly patch he had cultivated might fit into some larger scheme of things. And because so few scholars dare to explore beyond the boundaries of their own narrow fields, Dr. Wilson has produced an original work of synthesis time after time.

The Hungarian-American psychologist Mihaly Csikszentmihalyi examines the lifelong creative growth of a number of scientists, Wilson among them, in his 1996 book *Creativity: The Psychology of Discovery and Invention,* for which he interviewed Wilson at length. That was around the time when Wilson was writing *Consilience,* still focused on the problem of how to unify science and the humanities, but now without the condescension that had marred his introduction of sociobiology. "I see a picture forming," he told Csikszentmihalyi, "one in which I would pay a great deal more attention to the fundamentals of the social sciences." He hoped, he said, to use "the biologist's approach" to "winnow and reanneal the elements of the social sciences that I think are required to create a consilience" between those fields and biology.

From that beginning, Csikszentmihalyi moved on to ask about Wilson's development as a scientist. Wilson responded frankly, acknowledging two fundamental motivations: his love of his field—"I could happily spend 360 out of 365 days away from other people, you know, traveling in the rain forest and in my library"—and his lifetime of driving ambition. "The other thing," he said of that motivation, "is insecurity, ambition, a desire to control. A scientist—and this is a risky thing for me to confess—wishes to control, and the way to control is to create knowledge and have ownership of it."

At that time in his life—on the verge of retiring from Harvard, a significant transition—Wilson attributed his lifelong growth primarily to that desire to control:

> I want to feel that I'm in control, that I cannot be driven out of it, that I cannot be stopped, that I will be well regarded for being in it, and that entails control, and control means ambition. It means constantly extending one's reach, renewing, extending, innovating.

Csikszentmihalyi comments, "These were what current psychological jargon calls deficit motives, based on efforts to compensate for undesirable early experiences." Wilson's love of nature is of course the other engine of that drive, as he acknowledged to Csikszentmihalyi after speaking of control: "I think that the combination of those drives is what makes a major scientist."

A decade later, the balance had shifted; Wilson's confidence in himself had deepened profoundly. Retiring from Harvard in 1996 meant time freed up for writing, and from that year to 2020, Wilson poured forth his many books, all of them popular, several of them *New York Times* best-sellers. He even, for the fun of it, wrote a novel, *Anthill,* published in 2010, about a boy who finds a way to help save a stretch of Alabama wilderness from developers, complete with a fable of life in a colony of talking ants and a knock-down, drag-out fistfight that he told me he loved writing. His books and prize money made him moderately wealthy, enough to support living at Brookhaven, where Irene was now under extended care, and also to endow a chair at Harvard and a program for graduate students.

Yet he has hardly slowed down. The COVID pandemic confined him to Brookhaven; he studied alone in his apartment until three each afternoon, when he went to sit with Irene, working those 360 days out of 365. He was assembling another grand synthesis, the grandest one yet: pulling together the immense but scattered body of research on ecosystems to craft a general theory. "I've gone through the past ten years of papers in *Nature* and *Science,*" he told me during one of our regular phone calls, "tearing out any research reports that seem relevant and stacking them on my desk." He would mine other, more specialized journals as well. Then he would sit at his desk and read through them, make notes in his small cursive on yellow tablets, pages and pages, listen to music, and think, and find patterns, and write.

One day, before the pandemic locked us all away from one another, I was sitting with Ed in the Brookhaven garden room, he on a couch with his back to the courtyard windows, I in an armchair opposite him. His hair had long ago turned pure white; he still had it cut and combed it over boyishly, a rebellious lock straying forward at his right temple. Tall, lean, with long, thin legs, tanned, nearly indefatigable at the age of ninety, he might be mistaken for Uncle Sam if you missed the warmth and intelligence in his expression. The sun from the courtyard made a nimbus around his shoulders and his head.

How the topic came up I don't recall. We were probably talking about his ecosystem studies, since they were foremost in his mind. Ten years of *Nature* and *Science* is a mighty stack of weekly journals. In any case, Ed reached his arm behind him, fished in his back pocket, and pulled out a neatly folded white handkerchief. "Look," he said.

My recorder wasn't running—I'm terrible with recording equipment—so I can't set down an exact transcription here. But I remember what happened, remember the gist of what he said, and it was the most extraordinary demonstration I've ever seen in my life.

Ed shook out his handkerchief, making a flat white sheet. He might have been demonstrating what happens when an island sterilized by a volcanic eruption begins to reacquire living organisms. Or he might have been demonstrating what happened three billion years ago, when the first organisms came somehow alive in the shallows of the early Earth. "I've been studying origami," he said. "Paper folding." Then he said—now I'm paraphrasing—"The first organism to come ashore has the entire environment to itself. But then"—pinching up a corner of the handkerchief—"another organism arrives. They have to relate to each other, accommodate each other." More pinching, more folding. "Then another"—folding—"and another"—folding, harder to hold now as the peaks and valleys of folds begin to multiply—"and another. Complication builds on complication, sorting out territories, finding enough food, reproducing, limiting reproduction, changing behavior between enough space and overpopulation, evolving new organs and claiming new niches, predating, scavenging, thriving, going extinct." The flat handkerchief had become a small world of valleys, plains, peaks, ledges, draws. Filled with life. Nothing torn or broken. All of it connected. A flat sheet of energy folded into intricacy.

"I'm studying the mathematics of origami," Ed concluded. "I think it may allow me to model how ecosystems form."

And I thought then or later, probably later—I was much too dazzled in that sunny garden room with Ed Wilson folding his handkerchief into a world to think—Ed, too. From the blank sheet of his birth (ah, but with some genetic embroidery already worked onto it) through the launch of his childhood, the folds upon ever more intricate folds of his enlarging lifetime of work in science, its valleys, too, and ledges and peaks and draws. His life had been the model, the gestalt, of his scientific breakthroughs. Our bodies, our bodily minds, draw in the world through every pore. Intrude it inward into meaning. So he had done, and so he would continue to do, until his last breath.

"Well, let's go to lunch," Ed said. "Did you get what you needed?"

Acknowledgments

First and foremost, I owe a special debt of gratitude to the officers of the Alfred P. Sloan Foundation, Doron Weber in particular, for once again supporting my work. I hope it lives up to their high standards.

Edward O. Wilson made himself available for lengthy in-person interviews across the first six months of my research. Then the COVID pandemic prevented us from meeting in person, and we worked by phone. He was unfailingly open with information and generous with his time. We agreed that he should read the manuscript for factual accuracy, which he did. I was otherwise free of any limitations on what I might write; this is a collegial but not an authorized biography. Ed's assistant, Kathy Horton, facilitated the work, including sitting for an e-mail interview herself.

An old friend, the biographer and science journalist Victor McElheny, appeared one day in the hallway at Brookhaven, Ed's retirement community, and to my great surprise turned out to be a resident. Thus, conveniently, Vic gave me valuable insights into the life and work of Jim Watson, having written the man's biography. Small world.

Ed's former student Dan Simberloff, with a long and distinguished academic career himself, brought to life the adventure he shared with Ed of creating experimental Krakataus of Florida mangrove islands.

Without him, I would never have known about their mutual scientific work of chasing off sharks by knocking on their noses with rowboat oars.

Ryan Phelan and Stewart Brand helped me understand their pioneering effort to document the existence of every life-form on Earth, one of the several streams that flowed eventually into the mighty Encyclopedia of Life.

The Stanford University librarian Mike Keller has continued to make the facilities of the university library available to me, a service to scholarship and a great kindness. Jennifer Dill, sweet soul, kept me from sinking into lethargy.

Anne Sibbald, my longtime agent at Janklow & Nesbit Associates, negotiated arrangements with Doubleday with her usual wisdom and skill. Through all our years of working together, Anne has become a cherished friend. This was my second book with Gerry Howard at Doubleday—*Hedy's Folly* was the first—and, sadly, my last with Gerry, who retired at the end of 2020. Thomas Gebremedhin stepped in and stepped up to see *Scientist* through.

Ginger Rhodes read every chapter and advised me from her long experience as a clinical psychologist. My wife of more than thirty years, much beloved, she is also my best and closest friend.

Notes

N.B.: Quoted statements not referenced here derive from personal interviews.

1. SPECIMEN DAYS

2 “If a subject”: Wilson (2013), p. 45.
2 certified prodigy: ibid., pp. 24–25.
2 the first collector: Wilson (1994), p. 71.
3 “Take a subject”: Wilson (2013), pp. 45–46.
3 “Promethean intellect”: Wilson (1994), p. 44.
4 “The first time I ever”: EOW to IK, 23 March 1955. Wilson’s letters to his fiancée are unpublished.
5 “I am proud”: EOW to IK, 26 Nov. 1954.
6 “incredible paradise”: EOW to IK (postcard), 29 Nov. 1954.
6 “Never before or afterward”: Wilson (1994), pp. 165–66.
6 in lush forest: EOW to IK, 2 Dec. 1954.
6 “I am really”: ibid.
6 “human flesh”: EOW to IK, 3 Dec. 1954.
7 “It will prove at least”: *The Resolution Journal of Johann Reinhold Forster* (online).
7 “My work here”: EOW to IK, 6 Dec. 1954.
8 “It wasn’t too hard”: EOW to IK, 7 Dec. 1954.
8 “little patch of forest”: EOW to IK, 9 Dec. 1954.
8 “At the time I entered college”: Wilson (2013), p. 44.
9 “a stiff little hike”: EOW to IK, 13 Dec. 1954.

9 "It was truly": ibid.
9 "absolutely no signs": EOW to IK, 24 Dec. 1954.
9 "tried to climb a coconut tree": EOW to IK, 25 Dec. 1954.
9 "the strangest Christmas": EOW to IK, 24 Dec. 1954.
10 "colonial French": EOW to IK, 18 Dec. 1954.
10 a crowd of friends: EOW to IK, 24 Dec. 1954.
10 "curiosity and opportunism": Wilson (1994), p. 174.
10 "just about the most remote": EOW to IK, 28 Dec. 1954.
10 "Santo is the real": EOW to IK, 8 Jan. 1955.
10 "nothing more than": ibid.
10 "something out of a storybook": ibid.
11 "a betel-chewing": James A. Michener, "Return to the South Pacific," *Los Angeles Times,* 19 Oct. 1986 (online).
11 "an experience to be": EOW to IK, 8 Jan. 1955.
11 "*real* rainforest": ibid.
11 "picking up ants": EOW to IK, 10 Jan. 1955.
12 "Melanesian, as expected": Wilson (1994), p. 175.
12 "clobbered": EOW to IK, 12 Jan. 1955.
13 Wilson had helped clean: EOW to IK, 9 Jan. 1955.
13 "It was terribly rank": EOW to IK, 13 Jan. 1955.
13 "beautifully written": EOW to IK, 15 Jan. 1955.
13 "settled in a comfortable hotel": EOW to IK, 16 Jan. 1955.
13 "I mailed off": EOW to IK, 5 Jan. 1955.
13 "Seldom does a biologist": EOW to IK, 4 Jan. 1955.
14 "Ideas in science": Wilson (2013), p. 38.
14 "I have made friends": EOW to IK, 2 March 1955.
15 "the most primitive": EOW to IK, 21 Jan. 1955.
15 mentor Bill Brown: Brown and Wilson (1959), p. 29.
15 "the vast . . . Nullarbor Plain": EOW to IK, 18 Jan. 1955.
15 "a sunbaked little town": EOW to IK, 19 Jan. 1955.
15 *Nothomyrmecia* eluded them: Taylor (1978).
16 "What a country!": EOW to IK, 15 Feb. 1955.
16 "much more profitable": EOW to IK, 21 Feb. 1955.
17 "ceremonial scars": EOW to IK, 4 March 1955.
17 "the most exciting": EOW to IK, 6 March 1955.
17 1,000 per 100,000 population: Max Roser, "Ethnographic and Archaeological Evidence on Violent Deaths," in *Our World in Data* (online).
18 "final destination": EOW to IK, 9 March 1955.
18 "a cook-boy": ibid.
19 "just about the richest": EOW to IK, 11 March 1955.
19 "an expedition of": EOW to IK, 12 March 1955.
19 European population of twelve, "a new batch": EOW to IK, 31 March 1955.
19 "with 3 regular": EOW to IK, 4 April 1955.
20 "through the middle": EOW to IK, 31 March 1955.

20 "be acting as": EOW to IK, 2 April 1955.
20 "At the moment": EOW to IK, 3 April 1955.
20 "The forest floor": EOW to IK, 7 April 1955.
21 "One little boy": EOW to IK, 8 April 1955.
21 "There was much excitement": EOW to IK, 12 April 1955.
21 "The place has a miserable": EOW to IK, 14 April 1955.
21 "Happy Birthday!!": EOW to IK, 18 April 1955.
22 "even if you quit": quoted in EOW to IK, 23 April 1955.
22 a major paper in 1958: Wilson (1958a).
22 "the first white man": EOW to IK, 2 June 1955.
23 "The major part": EOW to IK, 4 June 1955.
23 five-week voyage: EOW to IK, 10 Aug. 1955; "This trip is fairly": EOW to IK, 23 July 1955.
23 "perhaps the most powerful": EOW to IK, 30 July 1955.
24 "a mild 'anti-intellectual' ": ibid.
24 specimens collected there: "Finished all my work this morning at the Paris Museum and enjoyed gazing at 160-year-old specimens of ants": EOW to IK, 26 Aug. 1955.
24 "The 1,000 species": EOW to IK, 30 Aug. 1955.
25 after 7 September: EOW to IK, 7 Sept. 1955, is addressed "London (still here!)." He was on "a day-to-day standby basis with various airlines," he reported. On 3 Sept. he had written that his chances were "very good" for the 9th. This sequence of letters ends here; in *Naturalist* (p. 200), Wilson recalls flying to New York on "September 5."
25 "Finally," he wrote: Wilson (1994), p. 200.
25 "a sanctuary and a realm": ibid., p. 56.

2. LOST WORLDS

27 "I stand in the shallows": Wilson (1994), pp. 5–6.
27 "It came into my world": ibid., p. 6.
27 "instinct of the knowledge": Herman Melville, *Moby-Dick,* chap. 42, "The Whiteness of the Whale" (online).
27 "all the mystery": Wilson (1994), p. 6.
27 "There was trouble": ibid.
28 "a favorite": Pinfish (Lagodon Rhomboides), Texas Parks and Wildlife Department pamphlet (online).
28 "It flew out": Wilson (1994), p. 13.
28 "Later," Wilson writes: ibid., pp. 13–14.
28 "terrifying nineteenth-century": ibid., p. 14.
28 "a carefully planned nightmare": ibid., p. 17.
28 "Though in many": Melville, *Moby-Dick* (online).
28 "how a naturalist": Wilson (1994), pp. 11–12.
29 "Pensacola, Mobile, Orlando": ibid., p. 52.

29 "A nomadic existence": ibid.
29 "the runt of my class": ibid., p. 40.
29 "just as happy": ibid., p. 56.
30 "It was my business": Doyle (1912, 1998), p. 21.
30 "big, metallic-colored": Wilson (2013), p. 95.
30 "Stalking Ants": Mann (1934).
30 "ants as savage": ibid., p. 171.
30 "led me to search": Wilson (2013), p. 96.
30 prospecting in faraway places: Wilson (1994), p. 149.
31 "large stretches of forest": Olmstead Brothers (1918), p. 1.
31 "Insects were everywhere": Wilson (1994), p. 58.
32 Wilson would report: Regnier and Wilson (1968).
32 "That day the little army": Wilson (1994), pp. 59–60.
32 "absorbed by the unending variety": ibid., pp. 56–57.
32 "My future was set": Wilson (2013), p. 98.
33 "snout butterflies": ibid.
33 "Big enough to hold": Wilson and Harris (2012), pp. 95–96.
34 "We visited my favorite": Wilson (1994), p. 71.
35 *Pheidole in the New World:* Wilson (2003b).
35 "the find of a lifetime": Wilson (2020), p. 51.
36 "Established colonies grew": ibid., p. 54.
36 "in effect a child workaholic": Wilson (1994), p. 71.
36 "All that I had become": ibid., p. 73.
37 "outdoor life": ibid., pp. 73–74.
37 "The Scout program": ibid., p. 75.
37 "Snake" Wilson; pygmy rattlesnake: ibid., pp. 77–79.
38 "the loss of a father": ibid., p. 42.
38 "deeply moved": ibid.
38 the Reverend Wallace: ibid., pp. 42–43.
38 "in a large tank": ibid., p. 43.
39 "paperboy": ibid., pp. 91–92.
40 "I liked their clean": ibid., p. 93.
40 "I was destined": ibid., p. 15.
41 "downward spiral": ibid., p. 41.
41 "I realized I could not depend": ibid., p. 97.
42 "So in June": ibid., p. 98.
42 volunteers with one blind eye: Wiltse (1967), p. 25.
42 "physical standards": Wilson (1994), p. 98.
42 "I vowed that although": ibid., pp. 98–99.
42 two thousand dollars: ibid., p. 101.
44 "The several best teachers": ibid., p. 109.
44 "On weekends and holidays": ibid., p. 113.
45 "into a jumble": ibid., p. 111.
45 "the sacred text": ibid., p. 110.

3. NATURAL SELECTION

46 "As many more individuals": Darwin (1859), p. 5.

47 "You have most cleverly": Darwin to T. H. Huxley, 25 Nov. 1859, *Darwin Correspondence Project* (online).

47 "the totally distinct problem": Huxley (1908), lecture six (online).

48 "the mental operations": Louis Agassiz (1874), *The Structure of Animal Life*, 3rd ed. (New York: Scribner, Armstrong), p. 118; *élan vital*: see Henri Bergson (1911), *Creative Evolution* (New York: Henry Holt), passim.

48 "I am strongly inclined": Darwin (1859), p. 8.

48 "Changes in the conditions": Winther (2001), p. 429.

48 "The direct action": Darwin (1868), vol. II, p. 472, quoted in Winther (2001), p. 447.

48 "Natural Selection": Darwin (1859), p. 5.

49 50 percent of Earth's heat: "Partial Radiogenic Heat Model for Earth Revealed by Geoneutrino measurements," *Nature Geoscience* 4 (Sept. 2011): 647–51.

49 Darwin answered Thompson: Darwin (1859), p. 282.

50 "I have been led": Darwin (1868), vol. II, p. 357.

50 "that the whole organisation": ibid., p. 358.

50 "Each cell or atom": Darwin to T. H. Huxley, 17 July 1865, *Darwin Correspondence Project* (online).

50 "circulate freely throughout": Darwin (1868), vol. II, p. 374.

51 writing in 1903: cited in Sandler and Sandler (1986), p. 755.

51 chromatic aberration: Bradbury (1989), p. 282.

51 "chromatin": Flemming (1880, 1965), p. 7.

51 "chromosome": Paweletz (2001), p. 74.

52 "Cell division involving": Flemming (1880, 1965), p. 9.

52 "mutual antagonism": Poulton (1894), pp. 126–27. Bowler (1983) usefully cites this discussion on his p. 77.

52 His "Experiments": Sandler and Sandler (1986), p. 754.

53 "the *transmission* of inherited": ibid., p. 755.

53 "Varieties enjoying even a slight": Larson (2004), p. 223.

54 "The particular evolutionary": Mayr and Provine, eds. (1980, 1998), p. 7.

54 "there was no doubt": ibid., p. 10.

54 "an appeal to experiment": Thomas Hunt Morgan, quoted in ibid.

55 "I learned to ask": ibid., pp. 205–6.

55 "I took leave": Wilson (1994), p. 116.

55 a new investigation: changed in 1971 (ibid., p. 117).

56 "my skills in the anatomy": ibid., p. 130.

56 "a callow, severely undereducated": Wilson (2013), p. 119.

56 "Although only seven years": ibid.

56 Like Wilson, Brown: William L. Brown, Cornell University Faculty Memorial Statement (online).

56 "was . . . one of the warmest": Wilson (1994), p. 132.
57 "long and deep": ibid.
57 "When I was sixteen": Lenfield (2011) (online).
58 ant caste system: "The Origin and Evolution of Polymorphism in Ants," reprinted in Wilson (2006c), pp. 19–41.
59 "sorrow and guilt-tinged": Wilson (1994), p. 125; "like a Bowery bum": ibid., p. 124; "I think he would": ibid., pp. 126–27.
59 "I was considered a prodigy": Wilson (2013), pp. 24–25.
60 "On a grander scale": Wilson (1994), p. 142.
60 "naturalist hobos": ibid., p. 143.
60 "My intellectual journey": Wilson (1992), p. 410.
61 "Even then, you could see": Claudia Dreifus, "A Conversation with E. O. Wilson," *New York Times*, 29 Feb. 2016.
61 "Nobody doubts": EOW to Irene Kelley, 28 March 1955.
62 "Stamp collectors": Wilson (1994), p. 219.

4. STAMP COLLECTORS AND FAST YOUNG GUNS

63 "I was among": Wilson (1994), p. 224.
63 "I was especially impressed": ibid., p. 44.
64 *What Is Life?:* Schrödinger (1944, 1967).
64 "was creating a sensation": Wilson (1994), p. 44.
64 "Many practitioners": Harwood (1994), p. 2; "exceptional individuals": Ernst Mayr, as paraphrased in Harwood (1994), p. 2.
64 they would thank: "The Nobel Prize in Physiology or Medicine 1962, Perspectives: What Is Life?" www.nobelprize.org (online).
65 "I got hooked": Watson (2017), p. 33.
65 "contain in some kind": Schrödinger (1944, 1967), p. 21.
66 "The other way": ibid., pp. 60–61.
66 "It was only later": Crick (1988), p. 18.
66 "When I arrived": Wilson (1994), p. 223.
67 "was then believed to be": Lederberg (1994), p. 424.
67 the phage group: Gunther S. Stent, "Introduction: Waiting for the Paradox," in Cairns, Stent, and Watson, eds. (1966), p. 5.
67 "an even speedier": Watson (2017), p. 38.
68 "like a hat in a hatbox": Watson, quoted in McElheny (2003), p. 280.
68 Hershey and Chase: Hershey and Chase (1982); Stahl (2001), p. 3.
69 "DNA was the hereditary": Linus Pauling: Watson (2017), p. 46.
69 "In *What Is Life?*": ibid., pp. 50–51.
70 "It has not escaped": Watson and Crick (1953).
70 "The science was now": Wilson (1994), p. 225.
70 "soon to receive": ibid., pp. 220–21.
71 "a confidence": Dietrich (1998), p. 85.
71 "On one occasion": Wilson (1994), p. 222.

71 "My self-esteem": ibid., p. 232.
71 "He treated most": ibid., p. 219.
71 "I found him": ibid.
73 "And Stanford would assist": ibid., p. 202.
74 "The traditionalists at Harvard": ibid., p. 221.
74 "had pulled off": ibid., p. 225.
74 "In examining every biological": Crick (1966), pp. 14–15.
75 "In 1960," he writes: Wilson (2013), p. 224.
75 "evolutionary biology" in 1958 course catalogue: Wilson (1994), p. 226.
75 " 'What shall we call' ": Wilson (2013), pp. 224–25.
76 "What we were now calling": ibid., p. 225.
76 "as able and ambitious": Wilson (1994), p. 232.
76 "I finally got around to calculus": ibid., p. 33.

5. SPEAKING PHEROMONE

79 neuropsychiatrist Auguste Forel: Parent (2003).
80 "It can therefore be": Forel (1908), p. 98.
80 "Pheromones," the zoologists wrote: Karlson and Lüscher (1959), p. 55.
80 "Humans communicate": Wilson (2020), p. 88.
81 "If I could find": ibid., p. 90.
81 "much like a moving pen": Wilson (1963), p. 103.
81 "the finest of all": Wilson (2020), p. 91.
82 "The pheromone might prove": ibid., p. 92.
82 "Barely visible": ibid., pp. 92–93.
83 "If successful, it would": ibid., p. 93.
83 three chemists: ibid., pp. 93–94.
84 "Mound nests of fire ants": ibid., p. 94.
84 "When the ants sense": Wilson (2013), p. 199.
84 By a rural Florida: Walsh et al. (1965), p. 321.
84 "Do what we can": Emerson (1865, 1876), p. 225.
84 "We came back [to Harvard]": Wilson (2013), p. 200.
85 "trail-following responses": Wilson (1959), p. 644.
85 "a relatively simple": Wilson (2013), p. 200.
86 "Was the pheromone": ibid.
86 The paper was sufficiently: Walsh et al. (1965).
86 "some frustration": ibid., p. 321.
86 "The advantage to the ants": Wilson (1963), pp. 105–6.
87 Not until 1981 did: Vander Meer et al. (1981).
87 "The trail substance": Wilson (2013), pp. 200–201.
87 "I was able," he recalled: Wilson, personal communication.
87 "a pheromone of sorts": Wilson (1963), p. 105.
88 "If live ants": Wilson (2020), p. 156.
88 "The removal of corpses": Wilson et al. (1958), p. 109.

88 For the first of: ibid., pp. 109–10.
88 Wilson varied the same: ibid., p. 110.
88 "smell faintly": Wilson (2020), pp. 156–57.
89 "There is no procedure": ibid., p. 157.
89 "The result": ibid.
90 "After being dumped": Wilson (1963), p. 105.
91 "A weak Spirit of Pismires": Wray (1670), p. 2065.
91 Vander Meer could summarize: Vander Meer (1983).
91 "put together pheromones": Wilson (2020), pp. 99–100.

6. KEYS

93 "By this time": Wilson (1994), p. 232.
93 1960 annual meeting: Programs planned for AAAS New York Meeting, *Science,* n.s. 132 (4 Nov. 1960), p. 1323.
94 "He's a real theoretician": Wilson (1994), p. 238.
94 "Robert (he resisted)": Wilson (2013), p. 220.
95 Its founder, Walter Hendricks: Toomey (2012).
95 A news story: *Time* 50 (8 Sept. 1947): 62.
95 Marlboro College became: "Marlboro Remembers John MacArthur," Marlboro College Web site, 27 Jan. 2017 (online); Jamie Harvey, Marlboro College database administrator, personal communication, 15 May 2020.
96 "In a committee decision": Kaspari (2008), p. 449.
96 "He would later tell": ibid.
96 niche theory, a concept: See MacArthur (1958) for a journal-length summary of the full dissertation.
96 redefining formally: Hutchinson (1957).
98 "The arguments MacArthur's": Kaspari (2008), p. 455.
98 *Darwin's Finches:* Lack (1947, 1983).
98 "For two days": Wilson (2006), p. 253.
99 "audacious and speculative": ibid., p. 254.
99 "I cannot speak": ibid.
99 "the growing boy": Sleigh (2007), p. 191.
99 influential 1937 book: Dobzhansky (1937).
99 "may seem provocative": Dobzhansky (1964), p. 443.
99 "Nothing succeeds": ibid.
100 "The opinion forcibly expressed": ibid., p. 445.
100 "A gene": ibid., p. 447.
101 "the molecular wars": Wilson (1994), p. 237.
102 "We both saw": Wilson (2013), p. 221.
102 "Theoretically," MacArthur writes: MacArthur (1958), p. 600.
102 "I spoke to MacArthur": Wilson (2013), pp. 225–26.
102 "These displayed in a simple form": ibid., p. 226.
103 "On small islands": ibid., p. 227.

103 Equilibrium model of fauna: MacArthur and Wilson (1963), p. 376.
104 "attracted great interest": McGuinness (1984), pp. 429–30.
104 "If most of the area": Diamond (1975), p. 129.
105 "We applied this simple": Wilson (2013), p. 227.
105 "The Genetical Evolution": Hamilton (1964).
105 "In certain circumstances": ibid., p. 1.
106 "To express the matter": ibid., p. 16.
107 "Half of Krakatau": MacArthur and Wilson (1963), p. 383.
109 "Breeding fauna of islands": Wilson and Simberloff (1969), pp. 269–70.
110 Tendrich ran field tests: ibid., p. 272.
113 thirty-foot steel tower: ibid., p. 276.
114 "dead individuals": ibid., p. 275.
114 "We report here": ibid., p. 278.
115 *The Theory of Island Biogeography:* Wilson and MacArthur (1967).
115 "Around the world": MacArthur and Wilson (1994), p. 256.
116 "The surgeon told him": Wilson (1994), p. 257.

7. FULL SWEEP

118 "the Caligula of biology": Wilson (1994), p. 219.
118 "The deciphering of the DNA": ibid., p. 223.
118 "I commented sourly": ibid., p. 224.
118 "He proved me wrong": ibid.
119 "Now I'm the head": Nan Robertson, "Love and Work Now Watson's Double Helix," *New York Times,* 26 Dec. 1980, p. A24.
119 "extended his hand": Wilson (1994), p. 222.
119 "In ten years": ibid., p. 224.
120 "Bethe wrote these articles": quoted in Bernstein (1980), pp. 44–45.
121 *Sphecomyrma* ("wasp ant") *freyi:* Wilson, Carpenter, and Brown (1967).
121 "truly intermediate": ibid., p. 10.
121 Bob Taylor, Wilson's: Taylor (1978).
121 "They are the first": Wilson, Carpenter, and Brown (1967), p. 17.
122 "the full sweep": Wilson (1971), p. 1.
122 "In most parts": ibid.
122 "i.e., nearly one": Darwin (1881), p. 144.
122 "50 grams of soil": Wilson (1971), p. 1.
123 Plants specialized to house ants: Hölldobler and Wilson (1990), pp. 534–46.
123 specialized food bodies: ibid., p. 547.
123 "turn and aerate": ibid., p. 549.
123 "exciting": Wilson (1971), p. 56.
124 "along the path": ibid.
124 "For an *Eciton*": T. C. Schneirla (1956), "The Army Ants," in *Report of the Smithsonian Institution for 1955,* pp. 379–406, quoted in Wilson (1971), p. 56.
124 "a kind of foundation": ibid., p. 58.

125 "Although the driver ant": Wilson (1971), pp. 71–72.
126 "Finally," Wilson concludes: Hölldobler and Wilson (1990), p. 55.
126 *Atta* colonies are huge: ibid., p. 51.
126 "A full-grown colony": ibid., p. 3.
126 "A rough calculation": ibid., p. 87.
127 "modified for riding": Wilson (1971), p. 27.
127 "Flying sperm dispensers": ibid., p. 157.
127 "Young workers tend": ibid., p. 163.
127 experiments Wilson describes: ibid., p. 165.
127 young workers usually: ibid., p. 163.
128 "neither exceptionally ingenious": ibid., p. 224.
128 "It usually results": ibid.
128 termite nest-building behavior: ibid., pp. 228–29.
129 "From these observations": ibid., p. 231.
130 Heylighen found only: Heylighen (2011–12), p. 2; by 2020: Google Scholar search.
130 "growing points": Polanyi (1962), p. 12.
130 "consists in the adjustment": ibid., p. 2.
131 "I was hardly": Wilson (1994), p. 309.
131 Cayo Santiago, just off: Kessler and Rawlins (2016).
131 "an intellectual turning point": Wilson (1994), p. 309.
132 "At once the number two": ibid.
132 "Primate troops and": ibid., p. 311.
133 "the systematic study": Wilson (1975), p. 4.
133 "populations follow at least": Wilson (1994), p. 312.
133 "one of the great": Wilson (1971), p. 458.

8. AMBIVALENCES

134 "the most important idea": Wilson (1994), p. 315. Wilson dates this encounter to the spring of 1965, but his passing Hamilton's paper along to Dan Simberloff confirms that he read it shortly after its publication in July 1964.
134 "I found an advantage": Wilson (1994), p. 319.
134 "The Genetical Evolution": Hamilton (1964).
134 "I was anxious": Wilson (1994), p. 319.
135 "We expect to find": Hamilton (1964), p. 16.
136 "Impossible, I thought": Wilson (1994), p. 319.
136 "As we departed southward": ibid., pp. 319–20.
136 "I was a convert": ibid., p. 320.
137 "in some respects": ibid.
137 "I thought I understood": ibid.
137 "One long, wild swing": Hamilton (1996), p. 25.
138 "wasn't a crank": ibid., p. 29.
138 "that I had somehow": quoted in Segerstråle (2013), p. 86.

138 "generally acceptable": Hamilton (1996), p. 29.
139 "I was traveling": ibid., pp. 29–30.
139 "I expected opposition": Wilson (1994), pp. 320–21.
140 "a discipline": Wilson (1978a), p. xi.
140 "led me to thinking": Dreifus (2019).
140 "the attempt," a reviewer: Caplan, ed. (1978), p. 3
140 "When the same parameters": Wilson (1971), p. 458.
140 "roused," as he admits: Wilson (1994), p. 323.
140 "In fact": ibid.
141 "Cathy," Wilson recalls: ibid., p. 324.
142 "Camus said": Wilson (1975), p. 3.
142 "That is wrong": ibid.
142 "The biologist, who is concerned": ibid.
143 "What, we are then compelled": ibid.
144 "But every time": *Odyssey*, book 11, line 593.
144 "arid" but "probably correct": Wilson (1974), p. 4.
144 "genetic consequences": ibid.
145 "It may not be too much": ibid.
145 "I believe that [sociobiology]": ibid.
146 There are an estimated: Wilson (2003b).
146 "All previous attempts": Wilson (1975), p. 16.
147 ten "qualities of sociality": ibid., pp. 16–19.
147 "The goal of investigation": ibid., p. 30.
148 "spent 2,900 hours": ibid., p. 31.
148 "centripetal movement": ibid., p. 38.
149 "No ordinary predator": ibid., p. 43.
149 "In some years": R. D. Alexander and T. E. Moore (1962), "The Evolutionary Relationships of 17-Year and 13-Year Cicadas, and Three New Species (Homoptera, Cicadidae, *Magicicada*)," *Miscellaneous Publications, Museum of Zoology, University of Michigan* 121: 39, quoted in Wilson (1975), pp. 42–43.
149 "The killer whales": Wilson (1975), p. 54.
150 "As air warms": fig. 3–14 in ibid., p. 61.
151 "It is a fact": ibid., p. 167.

9. HUMAN NATURES

152 "to consider man": Wilson (1974), p. 547.
152 "anatomically unique": ibid., pp. 547–48.
153 "a perceptive Martian": ibid., p. 548.
153 "mental hypertrophy": ibid.
154 "Our civilizations": ibid.
154 "Experience with other animals indicates": ibid.
154 "when mankind has achieved": ibid., pp. 574–75.
154 "The transition . . .": ibid., p. 575.

155 "[Sociobiology] will attempt": ibid.
155 "If the decision is taken": ibid.
155 "planned society . . .": ibid.
156 "In this, the ultimate": ibid.
156 "A world that can be explained": Albert Camus (1955), *The Myth of Sisyphus and Other Essays* (New York: Knopf), p. 6, quoted in Wilson (1975), p. 575.
156 "This, unfortunately": Wilson (1975), p. 575.
156 "has spread far": Boyce Rensberger, "Sociobiology: Updating Darwin on Behavior," *New York Times*, 28 May 1975, pp. 1, 52.
157 The Cambridge observers: Segerstråle (2000), pp. 18–19.
157 "There's no controversy yet": Segerstråle (2000), p. 18.
158 "equally impressive": Bonner (1975), p. 129.
158 Sociobiology Study Group was adding: Segerstråle (2000), pp. 19–20.
158 "screen of approval": quoted in ibid., p. 19.
159 "a political analysis": Heidi Benson, "New York Review of Books' Robert Silvers," *SF Gate*, 9 Nov. 2008 (online).
159 "the science concerned with": "epigenetic, *adj.* and *n.*," OED.
160 A letter of protest: *NYRB*, 13 Nov. 1975 (online).
160 "to say that word": Klehr (1988).
161 "*Now* there's controversy": Segerstråle (2000), p. 18.
161 "I thought there would be": Dreifus (2019) (online).
161 "struck by self-doubt": Wilson (1994), p. 339.
162 "ranked as coordinate": Wilson (1975), p. 4.
162 "Controversies involving sensitive": Segerstråle (2000), p. 15.
162 "Contrary to his critics' belief": ibid., pp. 39–40.
162 "I was not even an intellectual": Wilson (1994), p. 339.
163 "political upheaval": Segerstråle's words, paraphrasing Gould, in Segerstråle (2000), p. 24.
163 "Wilson," he told someone: quoted in Frankel (1979) (online).
163 had strongly supported: Wilson (1994), pp. 344–45; Segerstråle (2000), pp. 43–44.
163 "I was had": Segerstråle (2000), p. 44.
163 "In 1975": Wilson (1994), p. 339.
163 "I rethought my own": ibid.
164 "What has evolved": Wilson (1975), p. 559.
164 "the false statements": *NYRB*, 11 Dec. 1975 (online).
164 published the follow-up story: "The Politics in a Debate Over Sociobiology," *New York Times*, 9 Nov. 1975, p. 212.
165 "was reviled": Jumonville (2002), p. 587.
165 significant multicultural: Jumonville (2002), p. 587.
165 "Meta-analysis of the Heritability": Polderman, Benyamin, et al. (2015).
165 "a fresh surge": Wilson (1994), p. 339.
166 "not a work of science": Wilson (1978b), p. x.

10. THE DEEP THINGS

167 "great spiritual dilemmas": Wilson (1978b), pp. 2–3.
168 "SWEENEY:": Eliot (1932), pp. 24–25.
168 "Innate censors": Wilson (1978b), p. 5.
169 "soon," Wilson predicts: ibid.
169 "which of the [biological] censors": ibid., p. 6.
169 "The deep things": Oppenheimer (1964), p. 3.
169 "The only way forward": Wilson (1978b), pp. 6–7.
170 "every culture known": ibid., pp. 21–22.
170 "age-grading, athletic": ibid., pp. 22–23.
170 "is just one hodgepodge": ibid., p. 23.
171 "there will be no": ibid., p. 47.
171 "The predisposition": ibid., p. 169.
171 "According to the anthropologist": ibid.
171 "organize campaigns": ibid., pp. 170–71.
172 "by traditional methods": ibid., p. 172.
172 religion is unique to the human: ibid., p. 175.
172 "The votary is ready": ibid., p. 183.
172 "The extreme plasticity": ibid., pp. 185–86.
172 "The highest forms": ibid., p. 188.
173 "Marxism, traditional": ibid., p. 190.
173 "Marxism is sociobiology": ibid., p. 191.
173 "formulated as the enemy": ibid.
174 "I have wavered": Wade (1976), p. 1155.
174 "At times the protest": Wilson (2013), pp. 70–71.
174 This group, the International: Segerstråle (2000), pp. 21–22; Klehr (1988), p. 88.
174 "would most likely": Chagnon (2013), p. 382.
175 "because it laid a new": ibid., p. 383.
176 "He declares that": Richard D. Alexander, review of Sahlins (1976), in *American Anthropologist* 79 (1977): 917.
176 "We . . . indulge": Sahlins (1976), p. 10.
176 "The ballroom in which": Chagnon (2013), p. 384.
176 "Nobody heard me": ibid.
177 "I was nervous then": Wilson (1994), pp. 347–48.
177 "for his pioneering work": https://nationalmedals.org/laureate/edward-o-wilson.
178 "the general theme": Chagnon (2013), p. 385.
178 "hoping to put an end": ibid., pp. 385–86.
179 "that the first two": ibid., p. 386.
179 "and delivered from": ibid.
179 at a physical disadvantage: Wilson (1994), p. 348.
179 "Racist Wilson": Segerstråle (2000), p. 23. The group had used the same chant

against another academic, the political scientist Edward C. Banfield, at the University of Chicago in 1974. See "Protesters Disrupt Prof's U. of C. Talk," *Chicago Sun-Times*, 21 March 1974, p. 102.

179 "Several held up": Wilson (1994), p. 348.

179 "shouting 'Please stop!' ": Chagnon (2013), p. 387.

180 "Wilson, you're all wet!": Wilson (1994), p. 349.

180 "In a little over": ibid.

180 "The moderator then made": "Sociobiology Baptized as Issue by Activists," *Science* 199: 955.

180 "Of course they [applauded]": ibid.

11. CROSSING THE LINE

182 "the rich, natural": *New York Times Book Review*, 14 Jan. 1979, cited in "biophilia," *OED*.

183 "a reduction of habitat": Wilson (1994), p. 355.

183 "knotted with anxiety": ibid., p. 356.

183 "Speak too forcefully": ibid.

184 "which biological questions": Myers (1980), p. v.

184 "they teem with": Wilson (1994), p. 357.

184 Myers's grim conclusion: Myers (1999) (online).

184 "It is not unrealistic": Myers (1980), p. 15.

184 "A distinguished scientist": Wilson (1994), p. 357.

185 "Four [of the seven]": ibid., p. 354.

185 "Permit me to rephrase": "Resolutions for the 80s," *Harvard Magazine* 82, no. 3 (Jan.–Feb.): 21, reprinted in Wilson (2006b), p. 618.

185 "One day on impulse": Wilson (1994), p. 357.

186 Wilson and several of: "Conservation of Tropical Forests," *Science* 213 (Sept. 1981), p. 1314.

187 "the labyrinths of field biology": Wilson (1992), pp. 3–4.

188 "Even these species": ibid., p. 4.

189 "The forest at Angkor": Wilson (1988), p. 9.

189 "The 'SLOSS' debate": Thomas Lovejoy, Center for Biodiversity and Sustainability Web site, George Mason University (online).

189 "Fragment size also": Laurance et al. (2011), p. 58.

190 "The action": Wilson (1984), p. 25.

190 "The Brazilian Amazon": Lovejoy, Center for Biodiversity and Sustainability Web site (online).

190 The journal *BioScience:* Wilson (1985).

190 "largely inaccessible": ibid., p. 700.

191 "with 1 million": Stork (2018), p. 31.

191 "the natural human affections": Wilson (1985), p. 700.

191 "a molecule in a cloud": ibid., p. 701.

192 "why is there": ibid., p. 702.

192 "Only a tiny fraction": ibid.
193 "And in at least": ibid., p. 703.
194 "approximately equals": ibid., p. 704.
194 "The problems of Third World": ibid., pp. 704–5.
194 "To put the matter": ibid., p. 705.

12. REPRISING LINNAEUS

195 "featured more than": Wilson (1988), p. v.
195 "the number of species": ibid., pp. 13–14.
196 "It would be a great": ibid., p. 14.
196 "I will go further": ibid.
197 "the roughly 170,000": Raven and Wilson (1992), p. 1100.
197 "wandering around": quoted in David Pogue, "Pogue's Posts," *New York Times,* 23 Oct. 2008.
197 "Each species": Eisner, Lubchenco, Wilson, et al. (1995), p. 1231.
198 "I was taken by": personal communications, 9 Dec. 2020.
198 "For a good year": ibid.
199 "Comparative biology": Wilson (2003a), p. 77.
199 "At a deeper level": ibid., pp. 77–78.
200 "What we need": quoted in Pogue, "Pogue's Posts."
200 Wilson officially announced: www.ted.com/talks/e_o_wilson_my_wish (online).
202 "Our collaboration": quoted in Powell (2014) (online).
204 "We didn't write": quoted in ibid.
204 "The story of any": Tschinkel and Wilson (2014), p. 442.
204 "a wide-ranging subject": Wilson (1991), p. 13.
205 "The question I'm asked": ibid.
205 "currently studying": ibid., pp. 13–14.
205 "in the office": ibid., p. 14.
206 *Pheidole in the New World:* Wilson (2003b).
206 "Now, that might sound": quoted in Csikszentmihalyi (1996, 2013), pp. 273–74.
207 "perhaps fail to recognize": Segerstråle (2000), p. 314.
208 "One special difficulty": Charles Darwin, *On the Origin of Species,* quoted in Herbers (2009), p. 214.
208 "helping one's own": Wilson and Wilson (2007), p. 331.
209 "provide convincing evidence": ibid., p. 334.
209 "the fundamental problem": ibid., p. 328.
209 rarity of eusociality: Wilson and Hölldobler (2005b), p. 13369.
209 Another was the lack: ibid.
209 Finally, Hamilton's hypothesis: ibid.
210 "The Evolution of Eusociality": Nowak, Tarnita, and Wilson (2010).
210 "incendiary": Birch (2014), p. 381.
210 "sought out people": quoted in Johnson (2011) (online).

210 "is a mathematician": quoted in ibid.
210 "has shaken the very": Gintis (2012), p. 987.
211 Even Hamilton eventually realized: Wilson and Wilson (2007), p. 336.
211 "We hope our new theory": *Harvard Gazette*, 25 Aug. 2010.
211 "When Rabbi Hillel": Wilson and Wilson (2007), p. 345.
213 "that only by committing half": Wilson (2016), p. 3.

13. ORIGAMI

217 "I heard Wilson speak": Paul Simon, "The Year in Reading," *New York Times Book Review,* 19 Dec. 2016.
218 "In an alternative fate": Nicholas Wade, "Scientist at Work: From Ants to Ethics," *New York Times,* 12 May 1998, p. 74.
218 "I see a picture": quoted in Csikszentmihalyi (1996), pp. 266–67.
218 "I could happily": quoted in ibid., p. 269.
219 "I want to feel": quoted in ibid.
219 "These were what": ibid., p. 267.
219 "I think that": quoted in ibid., p. 269.

Bibliography

Albury, W. R. (1980). "Politics and Rhetoric in the Sociobiology Debate." *Social Studies of Science* 10: 519–36.

Alcock, John (2001). *The Triumph of Sociobiology.* Oxford, U.K.: Oxford University Press.

Altmann, Stuart A. (1962). "A Field Study of the Sociobiology of Rhesus Monkeys, *Macaca Mulatta.*" *Annals of the New York Academy of Sciences* 102: 338–435.

Ann Arbor Science for the People Editorial Collective (1977). *Biology as a Social Weapon.* Minneapolis: Burgess Publishing Company.

Arrhenius, Olof (1921). "Species and Area." *Journal of Ecology* 9, no. 1: 95–99.

Avery, Oswald T., Colin M. MacLeod, and Maclyn McCarty (1944). "Studies on the Chemical Nature of the Substance Inducing Transformation of Pneumococcal Types: Induction of Transformation by a Deoxyribonucleic Acid Fraction Isolated from Pneumococcus Type III." *Journal of Experimental Medicine* 79, no. 2: 137–58.

Axelrod, Robert, and William D. Hamilton (1981). "The Evolution of Cooperation." *Science* 211: 1390–96.

Ayala, Francisco J. (1985). *Theodosius Dobzhansky Biographical Memoir.* Washington, D.C.: National Academy of Sciences (online).

Bennett, Drake (2005). "The Evolutionary Revolutionary" [Robert Trivers]. *Boston Globe*, 27 March (online).

Berenbaum, May R. (2014). "Thomas Eisner 1929–2011." *Biographical Memoirs of the National Academy of Sciences.* Washington, D.C.: National Academy of Sciences.

Bernstein, Jeremy (1980). *Hans Bethe, Prophet of Energy.* New York: Basic Books.

Birch, Jonathan (2014). "Hamilton's Rule and Its Discontents." *British Journal of the Philosophy of Science* 65: 381–411.

Bonner, John Tyler (1975). Review of *Sociobiology: The New Synthesis*, by Edward O. Wilson. *Scientific American* 233, no. 4: 129–32.

Bowler, Peter J. (1983). *The Eclipse of Darwinism: Anti-Darwinian Evolution Theories in the Decades Around 1900*. Baltimore: Johns Hopkins University Press.

Bradbury, S. (1989). "Landmarks in Biological Light Microscopy." *Journal of Microscopy* 155, no. 3: 281–305.

Brown, Gerald E., and Sabine Lee (2009). "Hans Albrecht Bethe 1906–2005." *Biographical Memoirs of the National Academy of Sciences*. Washington, D.C.: National Academy of Sciences.

Brown, William L., Jr., and Edward O. Wilson (1959). "The Evolution of the Dacetine Ants." *Quarterly Review of Biology* 34: 278–94.

Burleigh, Michael (1994). *Death and Deliverance: 'Euthanasia' in Germany c. 1900–1945*. Cambridge, U.K.: Cambridge University Press.

Cairns, John, Gunther S. Stent, and James D. Watson, eds. (1966). *Phage and the Origins of Molecular Biology*. Cold Spring Harbor, N.Y.: Cold Spring Harbor Laboratory of Quantitative Biology.

Calhoun, John V. (2015). "Long-Lost Holotypes and Other Forgotten Treasures in the Ralph L. Chermock Collection, with Biographical Notes." *News of the Lepidopterists' Society* 57, no. 2: 80–85.

Caplan, Arthur L., ed. (1978). *The Sociobiology Debate: Readings on Ethical and Scientific Issues*. New York: Harper & Row.

Chadha, M. S., T. Eisner, et al. (1962). "Defence Mechanisms of the Arthropods—VII: Citronellal and Citral in the Mandibular Gland Secretion of the Ant *Acanthomyops Claviger* (Roger)." *Journal of Insect Physiology* 8, no. 2: 175–79.

Chagnon, Napoleon (2013). *Noble Savages: My Life Among Two Dangerous Tribes—the Yanomamö and the Anthropologists*. New York: Simon & Schuster.

Chagnon, Napoleon, and William Irons, eds. (1979). *Evolutionary Biology and Human Social Behavior: An Anthropological Perspective*. North Scituate, Mass.: Duxbury Press.

Cole, A. C. (1949). "The Ants of Bikini Atoll, Marshall Islands." *Pan-Pacific Entomologist* 25, no. 4: 172–74.

Cooper, Joseph B. (1985). "Comparative Psychology and Ethology." In Gregory A. Kimble and Kurt Schlesinger, eds., *Topics in the History of Psychology*, vol. 1. New York: Psychology Press.

Crick, Francis (1966). *Of Molecules and Men*. Seattle: University of Washington Press.

——— (1988). *What Mad Pursuit: A Personal View of Scientific Discovery*. New York: Basic Books.

Csikszentmihalyi, Mihaly (1996, 2013). *Creativity: The Psychology of Discovery and Invention*. New York: Harper Perennial Modern Classics.

Darlington, Philip J., Jr. (1957). *Zoogeography: The Geographical Distribution of Animals*. New York: John Wiley & Sons.

Darwin, Charles (1859). *On the Origin of Species by Means of Natural Selection, or the Preservation of Favoured Races in the Struggle for Life.* London: John Murray.

——— (1868). *The Variation of Animals and Plants Under Domestication,* 2 vols. London: John Murray.

——— (1881). *The Formation of Vegetable Mould Through the Action of Worms, with Observations on their Habits.* London: John Murray.

Diamond, Jared M. (1975). "The Island Dilemma: Lessons of Modern Biogeographic Studies for the Design of Natural Reserves." *Biological Conservation* 7: 129–46.

Dietrich, Michael R. (1998). "Paradox and Persuasion: Negotiating the Place of Molecular Evolution Within Evolutionary Biology." *Journal of the History of Biology* 31: 85–111.

Dinerstein, E., C. Vynne, A. R. Joshi, et al. (2019). "A Global Deal for Nature: Guiding Principles, Milestones, and Targets." *Science Advances* 5, no. 4: 1–17.

Dobzhansky, Theodosius (1937). *Genetics and the Origin of Species.* New York: Columbia University Press.

——— (1964). "Biology, Molecular and Organismic." *American Zoologist* 4, no. 4: 443–52.

Doyle, Arthur Conan (1912, 1998). *The Lost World.* Mineola, N.Y.: Dover.

Dreifus, Claudia (2019). "In Ecology Studies and Selfless Ants, He Finds Hope for the Future." *Quanta,* 15 May (online).

Duschinsky, Robert (2012). "*Tabula Rasa* and Human Nature." *Philosophy* 87: 509 29.

Eisenberg, J. F., and Wilton S. Dillon, eds. (1971). *Man and Beast: Comparative Social Behavior. Smithsonian Annual III.* Washington, D.C.: Smithsonian Institution Press.

Eisner, Thomas . . . Edward O. Wilson, et al. (1981). "Conservation of Tropical Forests." *Science* 213: 1314.

Eisner, Thomas, Jane Lubchenco, Edward O. Wilson, et al. (1995). "Building a Scientifically Sound Policy for Protecting Endangered Species." *Science* 269: 1231–32.

Eliot, T. S. (1932). *Sweeney Agonistes: Fragments of an Aristophanic Melodrama.* London: Faber & Faber.

Emerson, Ralph Waldo (1865, 1876). *Essays, First and Second Series.* Boston: Houghton Mifflin.

Emerson, Stephen A. (2014). *The Battle for Mozambique: The Frelimo-Renamo Struggle, 1977–1992.* Solihull, West Midlands, U.K.: Helion.

Fangliang He and Pierre Legendre (1996). "On Species-Area Relations." *American Naturalist* 148: 719–37.

Flemming, Walther (1880, 1965). "Contributions to the Knowledge of the Cell and Its Vital Processes." *Journal of Cell Biology* 25: 1–69.

Forel, Auguste (1908). *The Senses of Insects.* London: Methuen.

Forsyth, W. D. (1949). "The South Pacific Commission." *Far Eastern Survey* 18, no. 5: 56–58.

Frankel, Charles (1979). "Sociobiology and Its Critics." *Zygon* 15, no. 3: 255–73.

French, Howard W. (2011). "E. O. Wilson's Theory of Everything." *Atlantic* (Nov.) (online).

Fretwell, Stephen D. (1975). "The Impact of Robert MacArthur on Ecology." *Annual Review of Ecology and Systematics* 6 (Nov.): 1–13.

Futuyma, Douglas J. (1986). "Reflections on Reflections: Ecology and Evolutionary Biology." *Journal of the History of Biology* 19: 303–12.

Galton, Francis (1908). *Memories of My Life*. London: Methuen.

Gibson, Abraham H. (2012). "Edward O. Wilson and the Organicist Tradition." *Journal of the History of Biology* 46: 599–630.

Gintis, Herbert (2012). "Clash of the Titans" (review of Edward O. Wilson, *The Social Conquest of Earth*). *BioScience* 62: 987–91.

Giraldo, Ysabel Milton, and James F. A. Traniello (2014). "Worker Senescence and the Sociobiology of Aging in Ants." *Behavioral Ecology and Sociobiology* 68: 1901–19.

Gourevitch, Philip (2009). "The Monkey and the Fish." *The New Yorker*, 21 Dec., pp. 98–106, 108–11.

Gross, Paul R., and Norman Levitt (1994). *Higher Superstition: The Academic Left and Its Quarrels with Science*. Baltimore: Johns Hopkins University Press.

Hagen, Joel B. (1999). "Naturalists, Molecular Biologists, and the Challenges of Molecular Evolution." *Journal of the History of Biology* 32: 321–41.

Hamilton, W. D. (1963). "The Evolution of Altruistic Behavior." *American Naturalist* 97: 354–56.

——— (1964). "The Genetical Evolution of Social Behaviour." *Journal of Theoretical Biology* 7: 1–16.

——— (1996). *Narrow Roads of Gene Land: The Collected Papers of W. D. Hamilton. Vol. 1, Evolution of Social Behavior*. New York: Oxford University Press.

Harwood, Jonathan (1994). "Metaphysical Foundations of the Evolutionary Synthesis: A Historiographical Note." *Journal of the History of Biology* 27: 1–20.

Hendricks, Walter (1948). "Marlboro College." *Amherst Graduates' Quarterly* 37, no. 3: 181–87.

Herbers, Joan (2009). "Darwin's 'One Special Difficulty': Celebrating Darwin 200." *Biology Letters* 5, no. 2 (23 April): 214–17.

Hershey, A. D., and Martha Chase (1952). "Independent Functions of Viral Protein and Nucleic Acid in Growth of Bacteriophage." *Journal of General Physiology* 36: 39–56.

Heylighen, Francis (2011–12). "Stigmergy as a Generic Mechanism for Coordination: Definition, Varieties and Aspects." ECCO working paper.

Hogben, Lancelot (1943). *Mathematics for the Million*. New York: W. W. Norton.

Hölldobler, Bert, and Edward O. Wilson (1990). *The Ants*. Cambridge, Mass.: Belknap Press of Harvard University Press.

——— (2011). *The Leafcutter Ants: Civilization by Instinct*. New York: W. W. Norton.

Holmes, Samuel J. (1921). *The Trend of the Race: A Study of Present Tendencies in the Biological Development of Civilized Mankind*. New York: Harcourt, Brace.

——— (1939). "Darwinian Ethics and Its Practical Applications." *Science* 90: 117–23.

Holterhoff, Kate (2014). "The History and Reception of Charles Darwin's Hypothesis of Pangenesis." *Journal of the History of Biology* 47: 661–95.

Hutchinson, G. E. (1957). "Concluding Remarks." *Cold Spring Harbor Symposia on Quantitative Biology* 22: 415–27.

Huxley, Thomas Henry (1908). *Lectures and Essays.* New York: Cassell.

Iker, Sam (1982). "Islands of Life in a Forest Sea." *Mosaic* (Sept.–Oct.): 25–30.

Johnson, Jessica P. (2011). "Corina Tarnita: The Ant Mathematician." *The Scientist,* Sept. (online).

Jumonville, Neil (2002). "The Cultural Politics of the Sociobiology Debate." *Journal of the History of Biology* 35: 569–93.

Karlson P. and Lüscher A. (1959). "'Pheromones': A New Term for a Class of Biologically Active Substances." *Nature* 183: 55–56.

Kaspari, Michael (2008). "Knowing Your Warblers: Thoughts on the 50th Anniversary of MacArthur (1958)." *Bulletin of the Ecological Society of America* 89: 448–58.

Kay, Lily E. (2000). *Who Wrote the Book of Life? A History of the Genetic Code.* Stanford: Stanford University Press.

Kessler, Matthew J., and Richard G. Rawlins (2016). "A 75-Year Pictorial History of the Cayo Santiago Rhesus Monkey Colony." *American Journal of Primatology* 78: 6–43.

Kingsland, Sharon E. (1985). *Modeling Nature: Episodes in the History of Population Biology.* Chicago: University of Chicago Press.

Klehr, Harvey (1988). *Far Left of Center: The American Radical Left Today.* Piscataway, N. J.: Transaction.

Lack, David (1947, 1983). *Darwin's Finches.* Cambridge, U.K.: Cambridge University Press.

Larson, Edward J. (2004). *Evolution: The Remarkable History of a Scientific Theory.* New York: Modern Library.

Laurance, William F., and Richard O. Bierregaard, Jr. (1997). *Tropical Forest Remnants: Ecology, Management, and Conservation of Fragmented Communities.* Chicago: University of Chicago Press.

Laurance, William F., José L.C. Camargo, et al. (2011). "The Fate of Amazonian Forest Fragments: A 32-Year Investigation." *Biological Conservation* 144, no. 1: 56–67.

Lawler, Peter A. (2003). "The Rise and Fall of Sociobiology." *The New Atlantis* 1 (Spring): 101–12.

Lederberg, Joshua (1994). "The Transformation of Genetics by DNA: An Anniversary Celebration of Avery, MacLeod and McCarty (1944)." In James F. Crow and William F. Dove, eds., "Perspectives: Anecdotal, Historical and Critical Commentaries on Genetics." *Genetics* 136: 423–26.

Lehrer, Steven, ed. (2000). *Bring 'Em Back Alive: The Best of Frank Buck.* Lubbock: Texas Tech University Press.

Lenfield, Spencer Lee (2011). "Ants Through the Ages." *John Harvard's Journal,* July–August (online).

Levallois, Clement (2018). "The Development of Sociobiology in Relation to Animal Behavior Studies, 1946–1975." *Journal of the History of Biology* 51: 419–44.

Lewontin, R. C. (1991). *Biology as Ideology: The Doctrine of DNA*. New York: Harper Perennial.

Lewontin, R. C., Steven Rose, and Leon J. Kamin (1984). *Not in Our Genes: Biology, Ideology, and Human Nature.* 2nd ed. Chicago: Haymarket.

Lumsden, Charles J., and Edward O. Wilson (1981). *Genes, Mind, and Culture.* Cambridge, Mass.: Harvard University Press.

Lysenko, T. D. (1945). *Heredity and Its Variability.* Trans. Theodosius Dobzhansky. New York: King's Crown Press.

MacArthur, Robert H. (1958). "Population Ecology of Some Warblers of Northeastern Coniferous Forests." *Ecology* 39: 599–619.

——— (1972). *Geographical Ecology: Patterns in the Distribution of Species.* Princeton: Princeton University Press.

MacArthur, Robert H., and Edward O. Wilson (1963). "An Equilibrium Theory of Insular Zoogeography." *Evolution* 17: 373–87.

——— (1967). *The Theory of Island Biogeography.* Princeton: Princeton University Press.

Malcolm, Jay R. (1994). "Edge Effects in Central Amazonian Forest Fragments." *Ecology* 75: 2438–45.

Mann, W. M. (1934). "Stalking Ants, Savage and Civilized." *National Geographic,* Aug., pp. 171–92.

Maryanski, Alexandra (1994). "The Pursuit of Human Nature in Sociobiology and Evolutionary Sociology." *Sociological Perspectives* 37: 375–89.

Mayr, Ernst (1942, 1982). *Systematics and the Origin of Species.* New York: Columbia University Press.

——— (1959). "Where Are We?" *Cold Spring Harbor Symposia on Quantitative Biology* 24: 1–14.

——— (1982). *The Growth of Biological Thought: Diversity, Evolution, and Inheritance.* Cambridge, Mass.: Belknap Press of Harvard University Press.

Mayr, Ernst, and William B. Provine, eds. (1980, 1998). *The Evolutionary Synthesis: Perspectives on the Unification of Biology.* Cambridge, Mass.: Harvard University Press.

McElheny, Victor K. (2003). *Watson and DNA: Making a Scientific Revolution.* New York: Perseus Books.

McGuinness, Keith A. (1984). "Equations and Explanations in the Study of Species-Area Curves." *Biological Reviews* 59: 423–40.

McKie, Robin (2006). "The Ant King's Latest Mission." *The Guardian,* Oct. (online).

Melander, A. L., and F. M. Carpenter (1937). "William Morton Wheeler." *Annals of the Entomological Society of America* 30: 433–37.

Meselson, M., and F. W. Stahl (1958). "The Replication of DNA in Escherichia Coli." *Proceedings of the National Academy of Sciences USA* 44: 671–82.

Montagu, Ashley, ed. (1980). *Sociobiology Examined.* Oxford, U.K.: Oxford University Press.

Mora, Camilo, Derek P. Tittensor, Sina Adl, et al. (2011). "How Many Species Are There on Earth and in the Ocean?" *PLoS Biology* 9, no. 8.

Morange, Michel (1998). *A History of Molecular Biology.* Cambridge, Mass.: Harvard University Press.

Morgan, E. David (2008). "Chemical Sorcery for Sociality: Exocrine Secretions of Ants (Hymenoptera: Formicidae)." *Myrmecological News* 11: 79–90.

Myers, Norman (1980). *Conversion of Tropical Moist Forest: A Report Prepared by Norman Myers for the Committee on Research Priorities in Tropical Biology of the National Research Council.* Washington, D.C.: National Academy of Sciences.

——— (1999). *Conversations with History.* Institute of International Studies, UC Berkeley (online).

Nakamura, Jeanne, and Mihaly Csikszentmihalyi (2002). "The Concept of Flow." In C. R. Snyder & S. J. Lopez, eds. *Handbook of Positive Psychology.* New York: Oxford University Press, pp. 89–105.

Nowak, Martin A., Corina E. Tarnita, and Edward O. Wilson (2010). "The Evolution of Eusociality." *Nature* 466: 1057–62.

Oliveira, Paulo S. (1999). "Edward O. Wilson, Doyen of Biodiversity's Crusade, Honorary Fellow of ATB" [Association for Tropical Biology]. *Biotropica* 31: 538–39.

Olmstead Brothers (1918). *Rock Creek Park: A Report by the Olmstead Brothers.* Washington, D.C.: National Park Service (online).

Oppenheimer, J. Robert (1964). *The Flying Trapeze: Three Crises for Physicists.* London: Oxford University Press.

Oster, George F., and Edward O. Wilson (1978). *Caste and Ecology in the Social Insects.* Princeton: Princeton University Press.

Parent, André (2003). "Auguste Forel on Ants and Neurology." *Canadian Journal of Neurological Science* 30: 284–91.

Paul, Diane B. (1987). "'Our Load of Mutations' Revisited." *Journal of the History of Biology* 20: 321–35.

Paweletz, Neidhard (2001). "Walther Flemming: Pioneer of Mitosis Research." *Nature Reviews: Molecular Cell Biology* 2 (Jan.): 72–78.

Perry, John (1895). "On the Age of the Earth." Letter to the Editor. *Nature* 51: 341–42.

Pimm, Stuart L., Clinton N. Jenkins, and V. Li Binbin (2018). "How to Protect Half of Earth to Ensure It Protects Sufficient Biodiversity." *Science Advances* 2018, no. 4 (online).

Polanyi, Michael (1962). "The Republic of Science: Its Political and Economic Theory." *Minerva* 1, no. 1: 54–73.

Polderman, Tinca J. C., Beben Benyamin, Christiaan A. de Leeuw, et al. (2015). "Meta-analysis of the Heritability of Human Traits Based on Fifty Years of Twin Studies." *Nature Genetics* 47: 702–9.

Poulton, Edward B. (1894). In *Nature* 51: 126–27.

Powell, Alvin (2014). "Search Until You Find a Passion and Go All Out to Excel in Its Expression." *Harvard Gazette,* 15 April (online).

Pross, Addy (2012). *What Is Life? How Chemistry Becomes Biology.* Oxford, U.K.: Oxford University Press.

Ratowt, Sylwester Jan (2009). "Discordant Consensus: Dialogues on the Earth's Age in American Science, 1890–1930." Doctoral Dissertation, Department of the History of Science, University of Oklahoma Graduate College, Norman, Okla.

Raven, Peter H., and Edward O. Wilson (1992). "A Fifty-Year Plan for Biodiversity Surveys." *Science* 258: 1099–1100.

Regnier, F. E., and E. O. Wilson (1968). "The Alarm-Defense System of the Ant *Acanthomyops Claviger." Journal of Insect Physiology* 14: 995–70.

Ridley, Matt (1993). *The Red Queen: Sex and the Evolution of Human Nature*. New York: Harper Perennial.

Riskin, Jessica (2016). *The Restless Clock: A History of the Centuries-Long Argument over What Makes Living Things Tick*. Chicago: University of Chicago Press.

Robertson, Alan (1977). "Conrad Hal Waddington, 8 November 1905–26 September 1975." *Biographical Memoirs of the Fellows of the Royal Society* 23: 574–622.

Ross, Andrew (1993). "The Chicago Gangster Theory of Life." *Social Text* 35 (Summer): 93–112.

Ruse, Michael (2000). *The Evolution Wars: A Guide to the Debates*. New Brunswick, N.J.: Rutgers University Press.

Ruxton, Graeme, Stuart Humphries, Lesley J. Morrell, and David M. Wilkinson (2014). "Why Is Eusociality an Almost Exclusive Terrestrial Phenomenon?" *Journal of Animal Ecology* 83: 1248–55.

Sahlins, Marshall (1976). *The Use and Abuse of Biology: An Anthropological Critique of Sociobiology*. Ann Arbor: University of Michigan Press.

Sandler, Iris, and Laurence Sandler (1986). "On the Origin of Mendelian Genetics." *American Zoologist* 26(3): 753–68.

Schrödinger, Erwin (1944, 1967). *What Is Life? The Physical Aspect of the Living Cell*. Cambridge, U.K.: Cambridge University Press.

Schulman, Bruce J. (2001). *The Seventies: The Great Shift in American Culture, Society, and Politics*. New York: Da Capo Press.

Segerstråle, Ullica (2000). *Defenders of the Truth: The Battle for Science in the Sociobiology Debate and Beyond*. Oxford, U.K.: Oxford University Press.

——— (2013). *Nature's Oracle: The Life and Work of W. D. Hamilton*. New York: Oxford University Press.

Shipley, Brian C. (2001). "'Had Lord Kelvin a Right?': John Perry, Natural Selection and the Age of the Earth, 1894–1895." *Geological Society, London, Special Publications* 190: 91–105.

Simberloff, Daniel (1976a). "Experimental Zoogeography of Islands: Effects of Island Size." *Ecology* 57: 629–48.

——— (1976b). "Species Turnover and Equilibrium Island Biogeography." *Science* 194: 572–78.

Simberloff, Daniel S., and Lawrence G. Abele (1976). "Island Biogeography Theory and Conservation Practice." *Science* 191: 285–86.

Simberloff, Daniel S., and Edward O. Wilson (1970). "Experimental Zoogeography of Islands: A Two-Year Record of Colonization." *Ecology* 51: 934–37.

Skelly, David K., David M. Post, and Melinda D. Smith, eds. (2010). *The Art of Ecology: Writings of G. Evelyn Hutchinson.* New Haven: Yale University Press.

Slack, Nancy G. (2010). *G. Evelyn Hutchinson and the Invention of Modern Ecology.* New Haven: Yale University Press.

Sleigh, Charlotte (2007). *Six Legs Better: A Cultural History of Myrmecology.* Baltimore: Johns Hopkins University Press.

Slobodkin, L. B. (1993). "An Appreciation: George Evelyn Hutchinson." *Journal of Animal Ecology* 62: 390–94.

Smocovitis, Vassiliki Betty (1996). *Unifying Biology: The Evolutionary Synthesis and Evolutionary Biology.* Princeton: Princeton University Press.

Stahl, Franklin W. (2001). "Alfred Day Hershey 1908–1997." *Biographical Memoirs of the National Academy of Sciences* 80: 3–19.

Stork, Nigel E. (2018). "How Many Species of Insects and Other Terrestrial Arthropods Are There on Earth?" *Annual Review of Entomology* 63: 31–45.

Taylor, Robert W. (1978). "*Nothomyrmecia Macrops:* A Living-Fossil Ant Rediscovered." *Science* 201: 979–85.

Theraulaz, Guy, and Eric Bonabeau (1999). "A Brief History of Stigmergy." *Artificial Life* 5: 97–116.

Thomson, William (Lord Kelvin) (1862). "On the Age of the Sun's Heat." *Macmillan's Magazine* 5: 388–93.

Toomey, Daniel (2012). "'Believing It In': Robert Frost, Walter Hendricks, and the Creation of Marlboro College." *Robert Frost Review* 22 (Fall): 34–57.

——— (2013). "A Search for Patterns: The Life of Robert MacArthur." *Potash Hill,* Summer.

Trivers, Robert (2015). "Vignettes of Famous Evolutionary Biologists, Large and Small." *Unz Review,* 27 April (online).

——— (2015). *Wild Life: Adventures of an Evolutionary Biologist.* Boston: Plympton.

Tschinkel, Walter R., and Edward O. Wilson (2014). "Scientific Natural History: Telling the Epics of Nature." *BioScience* 64: 438–43.

Tyson, Charlie (2019). "A Legendary Scientist Sounds Off on the Trouble with STEM." *Chronicle of Higher Education,* 7 May.

UC San Diego (2002). "Proliferation of Argentine Ants in California Linked to Decline in Coastal Horned Lizards." *ScienceDaily,* 5 March.

Vandermeer, John H. (1972). "Niche Theory." *Annual Review of Ecological Systems* 3: 107–32.

Vander Meer, Robert K. (1983). "Semiochemicals and the Red Imported Fire Ant (*Solenopsis Invicta Buren*) (*Hymenoptera: Formicidae*)." *Florida Entomologist* 66, no. 1: 139–61.

Vander Meer, Robert K., Michael D. Breed, et al. (1998). *Pheromone Communication in Social Insects: Ants, Wasps, Bees, and Termites.* Boulder, Colo.: Westview Press.

Vander Meer, Robert K., F. D. Williams, and C. S. Lofgren (1981). "Hydrocarbon Components of the Trail Pheromone of the Red Imported Fire Ant, *Solenopsis Invicta.*" *Tetrahedron Letters* 22: 1651–54.

Volterra, Vito (1926). "Fluctuations in the Abundance of a Species Considered Mathematically." *Nature* 118: 558–60.

Wade, Nicholas (1976). "Sociobiology: Troubled Birth for New Discipline." *Science* 191: 1151–55.

Waller, John (2002). *The Discovery of the Germ: Twenty Years That Transformed the Way We Think About Disease.* New York: Columbia University Press.

Walsh, Christopher T., John H. Law, and Edward O. Wilson (1965). "Purification of the Fire Ant Trail Substance." *Nature* 207: 320–21.

Watson, F. E., and F. E. Lutz (1930). *Our Common Butterflies.* Guide Leaflet No. 38. New York: American Museum of Natural History.

Watson, James D. (1968). *The Double Helix: A Personal Account of the Discovery of the Structure of DNA.* New York: Touchstone.

——— (2001). *Genes, Girls, and Gamow: After the Double Helix.* New York: Alfred A. Knopf.

——— (2007). *Avoid Boring People: Lessons from a Life in Science.* New York: Alfred A. Knopf.

——— (2017). *DNA: The Story of the Genetic Revolution.* 2nd ed., rev. and updated. New York: Alfred A. Knopf.

Watson, James D., and Francis Crick (1953). "Molecular Structure of Nucleic Acids: A Structure for Deoxyribose Nucleic Acid." *Nature* 171: 737–38.

Weismann, August (1893). *The Germ-Plasm: A Theory of Heredity.* Trans. W. Newton Parker and Harriet Rönnfeldt. The Contemporary Science Series, ed. Havelock Ellis. London: Walter Scott.

Wetterer, James K., Alexander L. Wild, et al. (2009). "Worldwide Spread of the Argentine Ant, *Linepithema Humile.*" *Myrmecological News* 12: 187–94.

Wheeler, George C., and Jeanette Wheeler (1952). "The Ant Larvae of the Subfamily Ponerinae—Part II." *American Midland Naturalist* 48: 604–72.

Wheeler, William Morton (1910). *Ants: Their Structure, Development and Behavior.* New York: Columbia University Press.

Wild, Alexander L. (2004). "Taxonomy and Distribution of the Argentine Ant, *Linepithema Humile.*" *Annals of the Entomological Society of America* 97: 1204–15.

Wilson, David Sloan, and Edward O. Wilson (2007). "Rethinking the Theoretical Foundation of Sociobiology." *Quarterly Review of Biology* 82, no. 4: 327–48.

Wilson, Edward O. (1958a). "Patchy Distributions of Ant Species in New Guinea Rain Forests." *Psyche* 6, no. 1: 26–38.

——— (1958b). "A Chemical Releaser of Alarm and Digging Behavior in the Ant *Pogonomyrmex Badius* (Latreille)." *Psyche* 65, nos. 2–3: 41–51.

——— (1958c). "The Fire Ant." *Scientific American* 198, no. 3: 36–41.

——— (1959). "Source and Possible Nature of the Odor Trail of Fire Ants." *Science* 129: 643–44.

——— (1962). "Chemical Communication Among Workers of the Fire Ant *Solenopsis Saevissima* (Fr. Smith)." *Animal Behaviour* 10, nos. 1–2: 134–64.

——— (1963). "Pheromones." *Scientific American* 208, no. 5: 100–114.

——— (1971). *The Insect Societies.* Cambridge, Mass.: Belknap Press of Harvard University Press.

——— (1974). "The Conservation of Life." *Harvard Magazine* 76, no. 5: 28–37.

——— (1975). *Sociobiology: The New Synthesis.* Cambridge, Mass.: Belknap Press of Harvard University Press.

——— (1978a). "Foreword." In Arthur L. Caplan, ed., *The Sociobiology Debate: Readings on Ethical and Scientific Issues.* New York: Harper & Row.

——— (1978b). *On Human Nature.* Cambridge, Mass.: Harvard University Press.

——— (1978c). "What Is Sociobiology?" *Society,* Sept.–Oct., pp. 10–14.

——— (1980). "Resolutions for the 80s." *Harvard Magazine* 82, no. 3: 21.

——— (1984). *Biophilia.* Cambridge, Mass.: Harvard University Press.

——— (1985). "The Biological Diversity Crisis." *BioScience* 35: 700–706.

Wilson, Edward O., ed. (1988). *Biodiversity.* Washington, D.C.: National Academy Press.

——— (1991). "Ants." *Bulletin of the American Academy of Arts and Sciences* 45, no. 3 (Dec.): 13–23.

——— (1992). *The Diversity of Life.* Cambridge, Mass.: Belknap Press of Harvard University Press.

——— (1994). *Naturalist.* Washington, D.C.: Island Press.

——— (1996). "Macroscope: Scientists, Scholars, Knaves and Fools." *American Scientist* 86 (Jan.–Feb.): 6–7.

——— (2000). "A Memorial Tribute to William L. Brown (June 1, 1922–March 30, 1997)." *Psyche* 103, nos. 1–2: 49–53.

——— (2002). *The Future of Life.* New York: Alfred A. Knopf.

——— (2003a). "The Encyclopedia of Life." *Trends in Ecology and Evolution* 18, no. 2: 77–80.

——— (2003b). Pheidole *in the New World: A Dominant, Hyperdiverse Ant Genus.* Cambridge, Mass.: Harvard University Press.

——— (2006a). *The Creation: An Appeal to Save Life on Earth.* New York: W. W. Norton.

——— (2006b). *Nature Revealed: Selected Writings, 1949–2006.* Baltimore: Johns Hopkins University Press.

——— (2012). *The Social Conquest of Earth.* New York: Liveright.

——— (2013). *Letters to a Young Scientist.* New York: Liveright.

——— (2014). *A Window on Eternity: A Biologist's Walk Through Gorongosa National Park.* New York: Simon & Schuster.

——— (2016). *Half-Earth: Our Planet's Fight for Life.* New York: Liveright.

——— (2017). *The Origins of Creativity.* New York: Liveright.

——— (2020). *Tales from the Ant World.* New York: Liveright.

Wilson, Edward O., N. I. Durlach, and L. M. Roth (1958). "Chemical Releasers of Necrophoric Behavior in Ants." *Psyche* 65, no. 4: 108–14.

Wilson, Edward O., and M. S. Blum (1964). "The Anatomical Source of Trail Substances in Formicine Ants." *Psyche* 71, no. 1: 28–31.

Wilson, Edward O., Frank M. Carpenter, and William L. Brown, Jr. (1967). "The

First Mesozoic Ants, with the Description of a New Subfamily." *Psyche* 74, no. 1: 1–19.

Wilson, Edward O., and Robert H. MacArthur (1967). *The Theory of Island Biogeography.* Princeton: Princeton University Press.

Wilson, Edward O., and Daniel S. Simberloff (1969). "Experimental Zoogeography of Islands: The Colonization of Empty Islands." *Ecology* 50, no. 2: 278–96.

Wilson, Edward O., and William H. Bossert (1971). *A Primer of Population Biology.* Sunderland, Mass.: Sinauer Associates.

Wilson, Edward O., and G. Evelyn Hutchinson (1989). "Robert Helmer MacArthur 1930–1972." *Biographical Memoirs of the National Academy of Sciences.* Washington, D.C.: National Academy of Sciences.

Wilson, Edward O., and Bert Hölldobler (1994). *Journey to the Ants: A Story of Scientific Exploration.* Cambridge, Mass.: Belknap Press of Harvard University Press.

Wilson, Edward O., and Bert Hölldobler (2005a). "The Rise of the Ants: A Phylogenetic and Ecological Explanation." *Proceedings of the National Academy of Sciences* 102: 7411–14.

Wilson, Edward O., and Bert Hölldobler (2005b). "Eusociality: Origin and Consequences." Proceedings of the National Academy of Sciences 102(28): 13367–71.

Wilson, Edward O., and Jose M. Gomez Duran (2010). *Kingdom of Ants: Jose Celestino Mutis and the Dawn of Natural History in the New World.* Baltimore: Johns Hopkins University Press.

Wilson, Edward O., and Alex Harris (2012). *Why We Are Here: Mobile and the Spirit of a Southern City.* New York: Liveright.

Wiltse, Charles M., ed. (1967). *Physical Standards in World War II.* Washington, D.C.: Office of the Surgeon General, Department of the Army.

Winther, Rasmus G. (2001). "August Weismann on Germ-Plasm Variation." *Journal of the History of Biology* 34: 517–55.

Wray, John (1670). "Letter to the Editor." *Philosophical Transactions of the Royal Society of London* 68 (20 Feb.): 2064–66.

Wyhe, John van (2019). "Why There Was No 'Darwin's Bulldog': Thomas Henry Huxley's Famous Nickname." *The Linnean* 35, no. 1: 26–29.

Wylie, Philip (1942, 1983). *Generation of Vipers.* McLean, Ill.: Dalkey Archive Press.

——— (1947). *An Essay on Morals.* New York: Rinehart & Co.

Yudell, Michael, and Rob Desalle (2000). "Essay Review: Sociobiology, Twenty-Five Years Later." *Journal of the History of Biology* 33: 577–84.

Index

Page numbers in *italics* refer to illustrations.

Illustration Credits

In-text photos on pages 108, 111, and 112 are courtesy of E. O. Wilson.

FIRST INSERT

Page 1
Both images courtesy of E. O. Wilson

Page 2
Top: Creative Commons © Ajay Narendra, Samuel F. Reid, Chloe A. Raderschall
Bottom: Courtesy of E. O. Wilson

Page 3
Both images courtesy of E. O. Wilson

Page 4
Top: Courtesy of E. O. Wilson

Page 5
Top: Courtesy of the *American Journal of Primatology* 78(1): Special Issue: Cayo Santiago: 75 Years of Leadership in Translational Research. January 2016.

Page 6
Bottom: Wikimedia Commons © Turnbull SL

Page 7
Getty Images / LIFE Images Collection / Hugh Patrick Brown

Page 8
Top: Getty Images / LIFE Images Collection / Leonard McCombe
Bottom: Copyright © Yale University / William K. Sacco

SECOND INSERT

Page 1
Top: © Getty Images / LIFE Images Collection / Hugh Patrick Brown
Bottom: Courtesy of Marlboro College

Page 2
Top: © Sarah Blaffer Hrdy
Bottom: Courtesy Museum of Comparative Zoology, Harvard University.
© President and Fellows of Harvard College.

Page 3
Top: Office of Public Affairs, Yale University, Photographs of Individuals (RU 686).
Manuscripts and Archives, Yale University Library.

Page 4
Courtesy of E. O. Wilson

Page 5
Top: Courtesy of E. O. Wilson
Bottom: Courtesy of Harvard University

Page 6
Top: Ledgard, J. M. "Ants and us. Interview with E. O. Wilson."
The Economist (Intelligent Life) (Autumn), pp. 112–17.
Bottom: Courtesy of Dave Faulkner

Page 7
Both photos courtesy of E. O. Wilson

Page 8:
© E. O. Wilson Biodiversity Foundation

All other images either are provided by the author
or have an unknown photographer / copyright holder.

ALSO BY

RICHARD RHODES

WHY THEY KILL

The Discoveries of a Maverick Criminologist

Lonnie Athens was raised by a brutally domineering father. Defying all odds, Athens became a groundbreaking criminologist who turned his scholar's eye to the problem of why people become violent. After a decade of interviewing several hundred violent convicts—men and women of varied background and ethnicity, he discovered "violentization," the four-stage process by which almost any human being can evolve into someone who will assault, rape, or murder another human being. *Why They Kill* is a riveting biography of Athens and a judicious critique of his seminal work, as well as an unflinching investigation into the history of violence.

Current Affairs/Criminology

ALSO AVAILABLE

Arsenals of Folly
Hedy's Folly
John James Audubon
Masters of Death
The Twilight of the Bombs